Organic Chemistry: The Basics

Organic Chemistry: The Basics

Bradley Horne

Willford Press
www.willfordpress.com

Published by Willford Press,
118-35 Queens Blvd., Suite 400,
Forest Hills, NY 11375, USA

ISBN: 978-1-64728-005-5

Cataloging-in-Publication Data

Organic chemistry : the basics / Bradley Horne.
p. cm.
Includes bibliographical references and index.
ISBN 978-1-64728-005-5
1. Chemistry, Organic. 2. Chemistry. I. Horne, Bradley.
QD253.2 .O74 2022
547--dc23

For information on all Willford Press publications
visit our website at www.willfordpress.com

Willford Press

Contents

Preface

The compounds in which carbon forms a covalent bond with other atoms is called an organic compound. Organic chemistry is a sub-field of chemistry that focuses on the study of the properties, structure and reactions of organic compounds. The study of the properties of these compounds involves both physical and chemical properties. The study of the structure of organic compounds helps in determining their chemical composition and formula. There are various types of organic reactions which are studied within this field such as the chemical synthesis of natural products, polymers and drugs. Some of the chemicals which are dealt with in this field are hydrocarbons along with carbon bonded with other elements like oxygen, nitrogen, sulfur, phosphorus and halogens. This book is a compilation of chapters that discuss the most vital concepts in the field of organic chemistry. It provides significant information of this discipline to help develop a good understanding of this field as well as related fields. This textbook will serve as a valuable source of reference for those interested in this field.

Given below is the chapter wise description of the book:

Chapter 1- The branch of chemistry which deals with the study of the properties, structure, composition and reactions of carbon containing compounds is known as organic chemistry. The lasting attraction between atoms is known as a chemical bond. The chapter closely examines the key concepts of organic chemistry and chemical bonding to provide an extensive understanding of the subject.

Chapter 2- The chemical compounds which contain carbon are known as organic compounds. There are diverse types of organic compounds such as hydrocarbons, organometallic compounds, volatile organic compounds and semi-volatile organic compounds. This chapter discusses in detail these organic compounds and their properties.

Chapter 3- Reactions in organic chemistry are classified into a few broad categories. A few of these are acid-base reactions, addition reactions, substitution reactions and rearrangement reactions. The topics elaborated in this chapter will help in gaining a better perspective about these types of organic reactions.

Chapter 4- The sub-discipline of chemistry which studies the relative spatial arrangement of atoms in a molecule as well as their manipulation is known as stereochemistry. Chirality refers to a property of ions and molecules which are non-superimposable on their mirror images. Isomers are the molecules or ions which have the same chemical formula but different structures. The chapter closely examines these key concepts related to stereochemistry.

Chapter 5- Organic chemistry is used in a variety of different fields such as medicine and food industry. Some of the areas within food industry where organic chemistry is particularly used are coloring chemicals, food emulsifiers, flavor enhancing agents and moisture preservers. The diverse applications of organic chemistry in these fields have been thoroughly discussed in this chapter.

At the end, I would like to thank all those who dedicated their time and efforts for the successful completion of this book. I also wish to convey my gratitude towards my friends and family who supported me at every step.

Bradley Horne

1

Organic Chemistry: An Introduction

The branch of chemistry which deals with the study of the properties, structure, composition and reactions of carbon containing compounds is known as organic chemistry. The lasting attraction between atoms is known as a chemical bond. The chapter closely examines the key concepts of organic chemistry and chemical bonding to provide an extensive understanding of the subject.

Organic chemistry can be defined as the study of the structure, properties, and reactions of organic compounds and organic materials, i.e., matter in its various forms that contain carbon atoms. Carbon is unique in the variety and extent of structures that can result from the three-dimensional connections of its atoms.

The range of chemicals studied in organic chemistry include hydrocarbons, as well as myriad compositions based on carbon, but also containing other elements like oxygen, nitrogen, sulfur, phosphorus and the radiostable elements of the halogens.

A major focus of organic chemistry is the isolation, purification, and structural study of these naturally occurring substances. Many natural products are simple molecules. Examples include formic acid (HCO_2H) in ants, ethyl alcohol (C_2H_5OH) in fermenting fruit, and oxalic acid ($C_2H_2O_4$) in rhubarb leaves.

Chemists generally believed that compounds obtained from living organisms were endowed with a vital force that distinguished them from inorganic compounds. According to the concept of vitalism, organic matter was endowed with a "vital force". During the first half of the nineteenth century, some of the first systematic studies of organic compounds were reported.

The majority of chemical compounds occurring in biological organisms are in fact carbon compounds, so the association between organic chemistry and biochemistry.

The names of organic compounds are either systematic, following logically from a set of rules, or nonsystematic, following various traditions. Systematic nomenclature is stipulated by specifications from IUPAC.

Organic molecules are described more commonly by drawings or structural formulas, combinations of drawings and chemical symbols. The line-angle formula is simple and unambiguous. In this system, the endpoints and intersections of each line represent one carbon, and hydrogen atoms can either be notated explicitly or assumed to be present as implied by tetravalent carbon.

The concept of functional groups is central in organic chemistry, both as a means to classify structures and for predicting properties. A functional group is a molecular module, and the reactivity of that functional group is assumed, within limits, to be the same in a variety of molecules. Functional groups can have the decisive influence on the chemical and physical properties of organic compounds.

Molecules are classified on the basis of their functional groups. Alcohols, for example, all have the subunit C-O-H. All alcohols tend to be somewhat hydrophilic, usually form esters, and usually can be converted to the corresponding halides. Most functional groups feature heteroatoms (atoms other than C and H). Organic compounds are classified according to functional groups, alcohols, carboxylic acids, amines, etc.

The importance of organic chemistry in the present age is as immense as it had been since its inception. It plays an important role in our everyday life because food, medicines, paper, clothes, soap, perfumes, etc. are indispensable to us for proper living. The study of organic chemistry is important for chemists and pharmacists in synthesizing medicines for the alleviation of human suffering.

The reactions in organic chemistry occur between organic compounds. Let us now study about the different terminologies, classifications, field effects, types of reagents, the stability of intermediates, and properties in detail.

Cleavage of Bonds

The reactions in organic chemistry occur by the breaking and making of bonds. Bonds can cleave in either of two ways:

- Homolytic Cleavage
- Heterolytic cleavage

Homolytic Cleavage

If the covalent bonds between two elements break in such a way, that each of the elements gets their own electrons, it is called homolytic cleavage. That is, each element gets an electron. Homolytic cleavage results in the formation of free radicals.

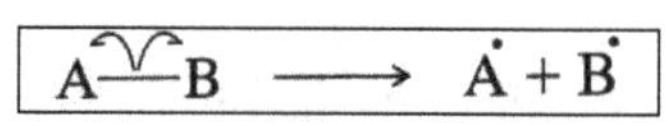

Homolytic Cleavage of Covalent Bond.

In the above figure, we have used an arrow to show the movement of electrons. Here, in this case, the arrow used is called the fish-hook arrow, as it signifies that there is a movement of only one electron.

Heterolytic Cleavage

If the covalent bonds between two elements break heterolytically, i.e., unequally, it results in the formation of charged species. This type of bond breaking, where the electrons are unevenly distributed is called heterolytic cleavage.

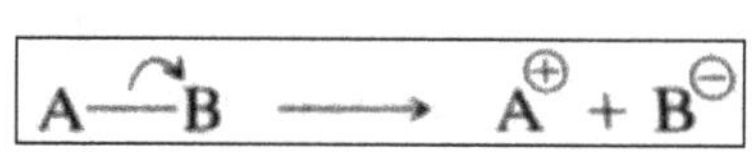

Heterolytic Cleavage of Covalent Bond.

In the above figure, we have used arrows to signify the movement of electrons, a regular arrow signifies that two electrons are being moved.

Reaction Intermediates in Organic Chemistry

Intermediates can be understood as the first product of a consecutive reaction. For example, in a chemical reaction, if $A \rightarrow B$ and $B \rightarrow C$, then, B can be said to be the intermediate for the reaction $A \rightarrow C$. The reactions in organic chemistry occur via the formation of these intermediates.

Carbenes

Carbenes (H_2C) are neutral and reactive species that have six electrons in the outer shell of carbon making them electron deficient. Since carbenes are species having two odd electrons, we can classify carbenes based on their spin states.

Singlet Carbene

The electrons are present in different orbitals with opposite spin. The electrons are paired in sp^2 hybridized orbitals and behave as paired electrons.

Spin state= $(2S + 1)$, S for singlet carbene is zero as the electrons are antiparallel.

Therefore, Spin state = $(2 \times 0 + 1) = 1$.

Triplet Carbene

Both the electrons are present in different orbitals and they possess the same spin.

Spin state = $(2S + 1)$, S for triplet carbene is 1 as both the electrons have the same spin.

Therefore, Spin state= $(2 \times 1 + 1) = 3$.

Hybridization of Singlet and Triplet Carbene

- Singlet Carbene Hybridisation: They are sp^2 hybridized with a bent shape. They have a bond angle of 103° and bond length of 112 pm.

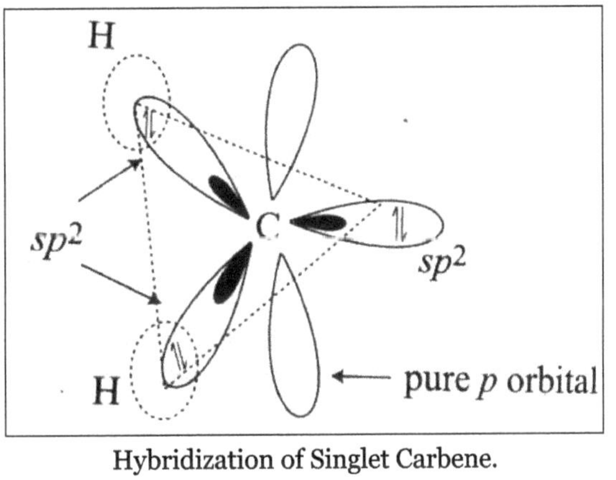

Hybridization of Singlet Carbene.

- Triplet Carbene Hybridization: They possess sp hybrid orbital with a linear shape. They have a bond angle and bond length of 180° and 103 pm respectively.

Stability of Singlet and Triplet Carbene

Triplet carbene has lower energy then singlet carbene because in singlet carbene there are more inter-electronic repulsions as both the electrons exist in the same orbital whereas in triplet carbene the two electrons exist in different orbitals making it possess less energy.

Free Radicals

Free radicals in organic chemistry are formed by the homolytic cleavage of carbon bond. The shape of the species formed is planar and the carbon is sp^3 hybridized with an odd electron being placed in the p-orbital. If the free radical is relatively stable then they may possess a planar structure.

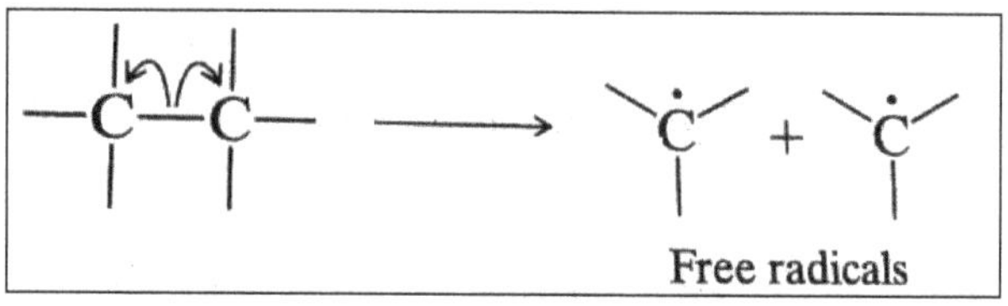

Planar Structure of Free Radicals.

Carbanions and Carbocations

Carbanions

They are generated by heterolytically cleaving a group attached to carbon without removing the bonded electrons. This makes the carbon have a pair of electrons, thereby, imparting a negative charge on the carbon. CH_3^- is isoelectronic with NH_3 and it is sp^3 hybridized and the shape is pyramidal owing to the presence of a lone pair of electrons.

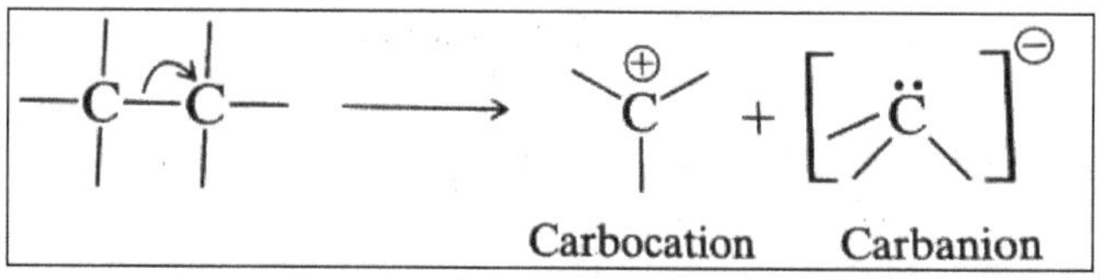

Formation of Carbanions and Carbocations.

Carbocations

Carbocations have a sextet of electrons on the carbon-containing the positive charge and hence termed 'cation'. It is sp^2 hybridized and has an empty p-orbital. The shape is planar. It is generally formed by heterolytic cleavage of a carbon-heteroatom bond.

Transition State in Organic Reactions

We saw the intermediates that could be formed in an organic reaction, now let us look into transition states and the difference between an intermediate and a transition state.

The intermediates in organic chemistry are formed in a multi-step reaction but some reactions can occur in a single step without having to form an intermediate. These reactions will occur by going through a transition state. This can be clear by looking at the energy profile diagram for a reaction, R→P.

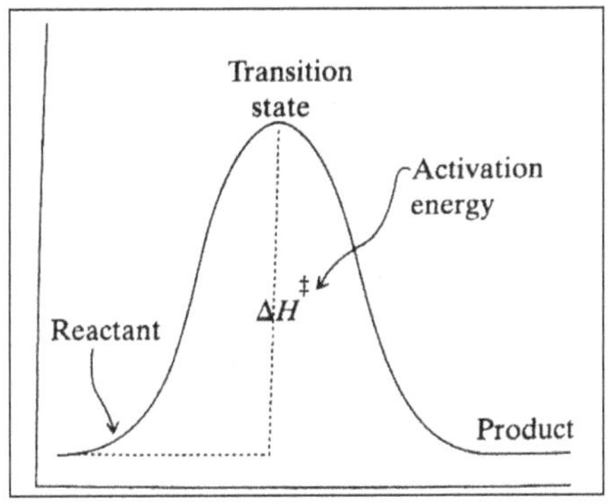

Energy diagram of single-step reaction.

The transition state corresponds to the highest energy in the reaction, after which it can give either the products or in the case of a reversible reaction, the reactants.

Consider a reaction, A→D with the following steps, $A \rightarrow B, B \rightarrow C,$ and $C \rightarrow D$.

The energy profile for this reaction is given below:

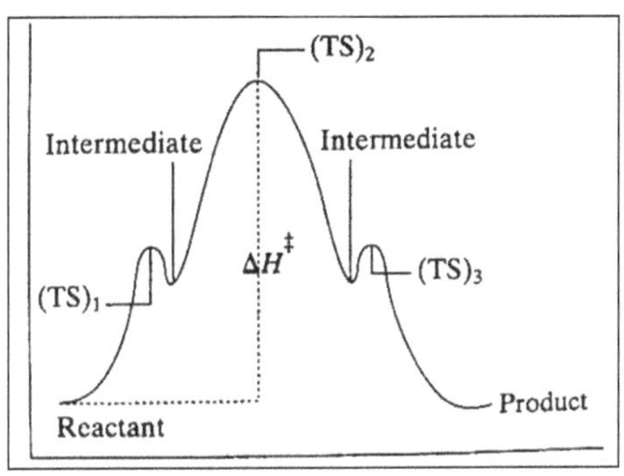

Energy profile of multiple-step reaction.

We can see that B and C are the products of a reaction and hence they are termed as intermediates. The highest energy of a particular reaction should be the transition state.

From the above example we can show that the intermediates are isolable, that is, they can be isolated. On the other hand, the transition state is not isolable because we assume the reaction to take place via a transition state cannot be isolated.

Reagents in Organic Chemistry

Reagents are the chemicals that we add to bring about a specific change to an organic molecule. Any general reaction in organic chemistry can be written as:

Substrate + Reagent → Product

Where the substrate is an organic molecule to which we add the reagent. Based on the ability to either donate or abstract electrons, the reagents can be classified as:

- Electrophiles
- Nucleophiles

Electrophiles

Electrophiles are electron deficient organic reagents. It can be generalized that all the positive charge containing species are electrophiles. For example, H^+, NO_2^+, CH_3^+, Cl^+.

Neutral molecules that are electron deficient can also act as electrophiles. Lewis acids like $AlCl_3$ and BF_3 are examples of neutral electrophiles.

Nucleophiles

Nucleophiles are electron rich organic reagents. They seek for bonding centers with other nuclei and hence the name, nucleophile. It can be generalized that negative charge containing species are nucleophiles. For example, H^-, CH_3^-, and Cl^-.

Neutral molecules with a lone-pair of electrons on the heteroatom can act as a nucleophile. For example, H_2O, NH_3, CH_3OH.

Types of Reactions in Organic Chemistry

Organic reactions are reactions that occur between organic compounds. The reactions in organic chemistry are broadly classified into six categories.

Substitution Reactions

$$R-X + Y \rightarrow R-Y + X$$

Where R-X is the substrate, Y is the reagent (which can be electrophilic or nucleophilic) and X is called the leaving group. The term substitution means one group is replacing the other group.

Types of Substitution Reaction:

1. Nucleophilic Substitution (S_N1, S_N2, S_Ni)
2. Electrophilic Substitution (S_E)
3. Nucleophilic Aromatic Substitution (S_NAr)
4. Addition Reactions

Addition reactions can be further classified into:

1. Electrophilic Addition
2. Nucleophilic Addition
3. Elimination reactions

Elimination Reactions

These reactions can be said to be the reverse of an addition reaction, wherein a simple molecule

(HX, H_2O) is removed from the substrate i.e. a molecule is said to be eliminated from the substrate. Elimination reaction can be classified further into E1, E2, E1CB.

1. Oxidation and reduction reactions,
2. Pericyclic reactions,
3. Molecular rearrangements.

Field Effect in Organic Chemistry

Inductive Effect

It is an electron delocalization effect via σ bonds that arises due to the difference in electronegativities. For example, in a σ bonded organic compound like C-C-C-Cl, the carbon attached to the chlorine atom can be referred to as the α-carbon, and the one adjacent that carbon as the ß -carbon and so on.

Now, since Chlorine is more electronegative than carbon, it withdraws the electrons that are present via the σ bond toward itself, thereby making $C\alpha$ fractionally positive. Since it is devoid of electrons, Cα now being slightly electropositive than $C\beta$ pulls the sigma bonded electrons of $C\alpha - C\beta$ bond toward itself and in this process, it makes $C\beta$ slightly electropositive.

The electron withdrawing effect of the Chlorine atom is being transmitted through the carbon chain via the σ bonds. This transmission of charges decreases rapidly with the number of intervening σ bonds. We can practically ignore this effect beyond $C\beta$.

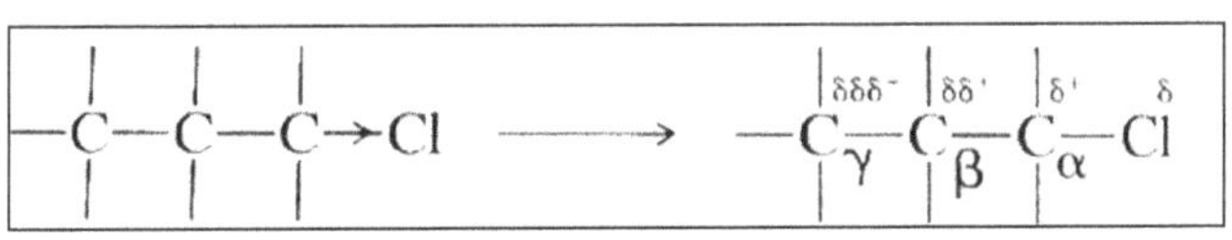

–I-effect

The arrow is pointed toward the more electronegative atom. If a group withdraws the electron from carbon, it makes carbon slightly electropositive. Such groups are called -I groups and the effect as -I effect. For example, $-Cl, -Br, -CN$ and $-NO_2$ are -I groups.

Groups that release electrons towards carbon are termed as +I groups and the effect is termed as +I inductive effect in organic chemistry. For example, alkyl groups like $-CH_3$ are +I groups.

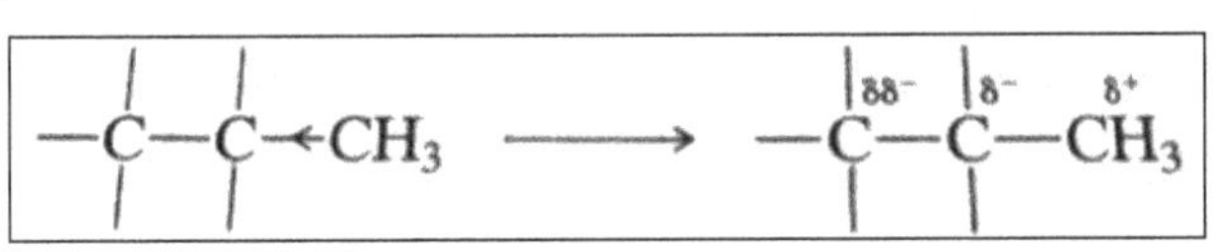

+I-effect

Electromeric Effect

It is the temporary delocalization of π-electrons in a compound containing multiple covalent bonds. It is important to note that it is only a temporary effect, that is, it occurs only when a reagent is added. Electromeric Effect in organic chemistry can be classified into two types:

- Positive Electromeric Effect
- Negative Electromeric Effect

Positive Electromeric Effect

When the π-electrons are given to the attacking reagent. For example, the reactions alkenes and alkynes mostly occur via +E, this reaction is also called electrophilic addition.

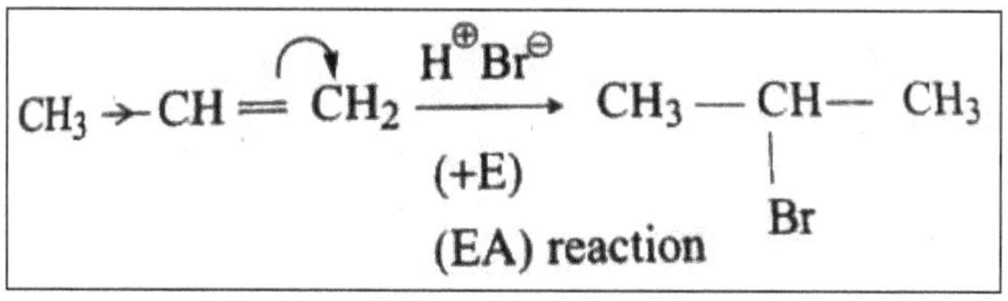

Positive Electromeric Effect.

The proton adds at C-1 as the π-electrons were given to the attacking reagent (H^+). This results in the formation of a carbocation.

Negative Electromeric Effect

When the π-electrons are shifted to a more electronegative atom (O, N, S) joined via multiple bonds. For example, the reactions of aldehydes and ketones occur predominantly by -E effect. It is also called nucleophilic addition.

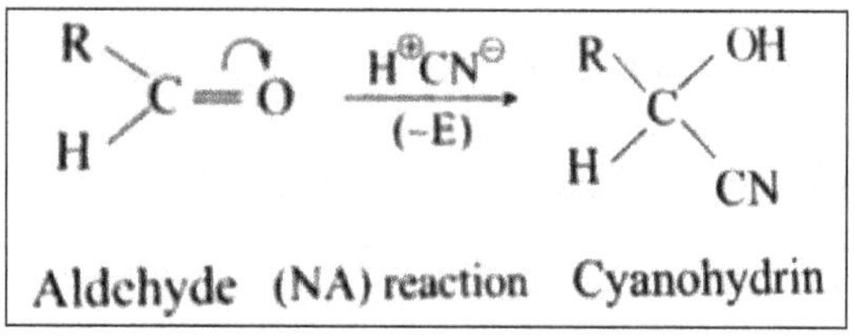

Negative Electromeric Effect.

The CN- ion adds to the C atom of the carboxy group opposite to the movement of the π-electron cloud.

Mesomeric Effect

Molecules possessing sigma-bonds and pi-bonds alternatively exhibit the mesomeric effect. The effect is exhibited due to the permanent delocalization of π-bonds. This increases the number of resonating structures which makes the molecules of organic chemistry more stable. Such kind of a system where there are alternative sigma and pie bonds are called conjugated.

Types of Mesomeric Effect

- Positive mesomeric effect
- Negative mesomeric effect

Positive Mesomeric Effect

This effect is exhibited when the direction of the delocalization of electrons is away from the position where the group is attached. Normally groups having a lone pair of electrons attached to a conjugated system push electrons into the conjugated system, that is, away from them.

Groups in organic chemistry showing positive mesomeric effect (+M) effect are -OH, -OR, $-NH_2$, -SH, -X, etc.

Negative Mesomeric Effect

This effect is exhibited when the direction of the delocalization of electrons is towards the position where the group is attached. These are generally electron withdrawing groups of organic chemistry.

Resonance Effect

For certain molecules like carbonate ion (CO_3^{2-}), one single Lewis structure would not be enough to explain all of the properties. In that case, the molecule is said to have more than one structure.

Each of those structures can explain some of the properties but not all of the properties. The actual structure of the molecule is a hybrid of all the possible structures (canonical forms). This phenomenon is called resonance in organic chemistry. If resonance occurs, each bond would be both, a single bond and a double bond at the same time i.e. the bond order would lie between one and two.

Resonating structures should fulfill the following criteria:

1. All atoms should have the same positions in all the structures.
2. There should have the same number of paired and unpaired electrons.
3. The structures should have almost the same energies.

Canonical forms do not have any existence in reality.

Resonance Energy

The energy difference between the most stable canonical form and the resonance hybrid is known as resonance energy. More the resonance energy, more is the stability.

Rules for finding out the most stable canonical form:

1. The canonical form with no charges is the most stable.
2. The canonical form with more number of covalent bonds is more stable.
3. The canonical form where unlike charges are in close proximity are more stable.
4. If there are to be charged, the negative charge should be on an electronegative atom. Then this canonical form is said to be more stable.

Role of Steric Hindrance

The structure and reactivity of many compounds in organic chemistry are greatly dictated by the presence of bulky groups or constituents in the molecule. This is called steric hindrance. It arises because of inter-electronic repulsions due to spatial crowding amongst bulky groups. Using steric factors, we can conclude that trans-2-butene is more stable than cis-2-butene.

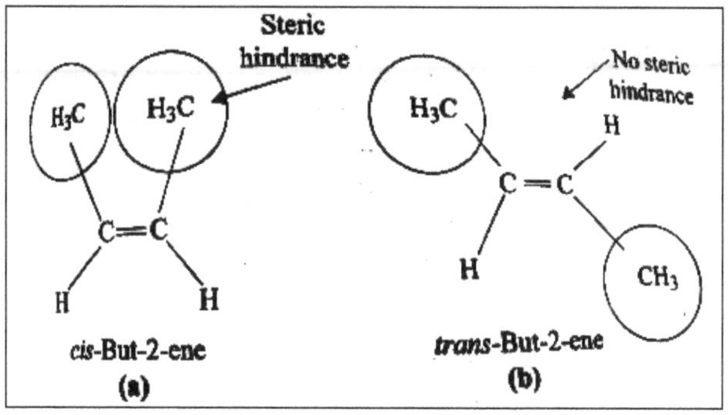

Steric Hindrance.

Stability of Intermediates

Carbocations

The stability of carbocations can be explained by the Inductive Effect.

$CH_3^{+}\,^{+}CH_2(CH_3)\ ^{+}CH(CH_3)_2\,^{+}C(CH_3)_3$ i.e. Methyl Primary (1°) Secondary (2°) Tertiary(3°).

We know that alkyl groups of organic chemistry are +I groups, that is, they release electrons through the sigma bonds.

Since the carbon is deficient of electrons, we can say that as the number of methyl group increases, the stability of the carbocation increases. Since the electropositive carbon is satiated by the electrons given by the methyl groups via +I effect.

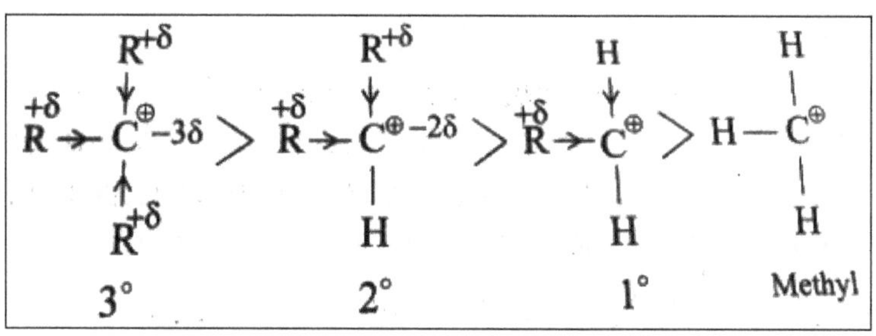

Therefore, the stability order of carbocations in organic chemistry is of the order 3° > 2° > 1° > Methyl carbocation.

Hyperconjugation

When a C-H σ-bond is in conjugation with a carbocation, this effect is observed. A carbocation has a vacant p-orbital. The bonded σ-electron pair of the C-H bond is displaced toward the vacant p-atomic orbital. This increases the electron density in the empty p-AO.

It is, therefore, a resonance effect where a C-H bond breaks and the σ-electron pair is delocalized to the vacant p-AO of the carbonation. Since the bond between C and H is broken, it is also called as 'no bond resonance'. It is also referred to as the Baker-Nathan effect.

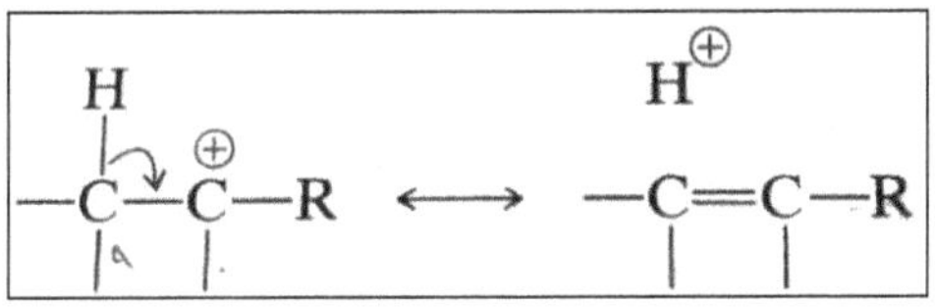

More the number of α-hydrogens in a carbocation, that many hyperconjugated structures would be possible. More the number of structures more is the stability of the species.

Stability of Carbanions

The stability of carbanions can be explained using the inductive effect.

$CH_3 - CH_2(CH_3) - CH(CH_3)_2 - C(CH_3)_3$ i.e. Methyl, Primary (1°), Secondary (2°), Tertiary (3°).

Since alkyl groups are electron releasing by nature through induction, we can say that more the number of methyl groups attached to the carbon having a negative charge, less would be its stability.

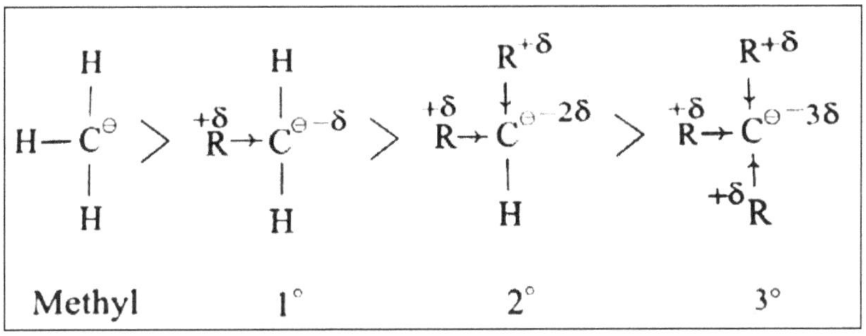

This is because, the carbon already has a negative charge, to which the methyl groups push electrons via induction. This results in inter-electronic repulsions and destabilizes the species.

Therefore the stability order for carbanions is as follows: 3°< 2°< 1°< Methyl carbanion.

Stability of Free Radicals

The stability of free radicals in organic chemistry follows the same trend as that of carbocations.

$CH_3–CH_2(CH_3)–CH(CH_3)_2–C(CH_3)_3$ i.e. Methyl, Primary (1°), Secondary (2°), Tertiary(3°).

Therefore, the stability order of free radicals is of order: 3° > 2° > 1° > Methyl carbocation.

This can be explained with the help of Hyperconjugation that we saw, but here there would be an overlap between the σ-bond of C-H and the odd electron in the p-orbital of carbon.

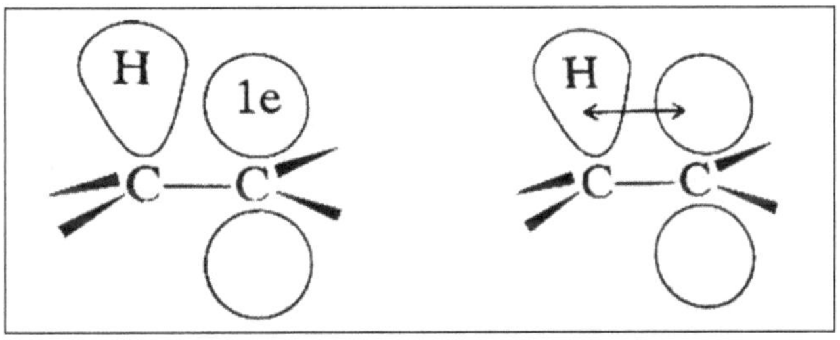

Chemical Bonding

Chemical bonding is one of the most basic fundamentals of chemistry that explains other concepts such as molecules and reactions. Atoms are attracted to each other or how products are formed after a chemical reaction has taken place. To understand the concept of bonding, one must first know the basics behind atomic structure.

A common atom contains a nucleus composed of protons and neutrons, with electrons in certain energy levels revolving around the nucleus. The main focus will be on these electrons. Elements are distinguishable from each other due to their "electron cloud," or the area where electrons move around the nucleus of an atom. Because each element has a distinct electron cloud, this determines their chemical properties as well as the extent of their reactivity (i.e. noble gases are inert/not reactive while alkaline metals are highly reactive). In chemical bonding, only valence electrons, electrons located in the orbitals of the outermost energy level (valence shell) of an element, are involved.

Lewis Diagrams

Lewis diagrams are graphical representations of elements and their valence electrons. Valence electrons are the electrons that form the outermost shell of an atom. In a Lewis diagram of an element, the symbol of the element is written in the center and the valence electrons are drawn around it as dots. The position of the valence electrons drawn is unimportant. However, the general convention is to start from 12 o'clock position and go clockwise direction to 3 o'clock, 6 o'clock, 9 o'clock, and back to 12 o'clock positions respectively. Generally the Roman numeral of the group corresponds with the number of valence electrons of the element.

Below is the periodic table representation of the number of valence electrons. The alkali metals of Group IA have one valence electron, the alkaline-earth metals of Group IIA have 2 valence electrons, Group IIIA has 3 valence electrons, and so on. The nonindicated transition metals, lanthanoids, and actinoids are more difficult in terms of distinguishing the number of valence electrons they have.

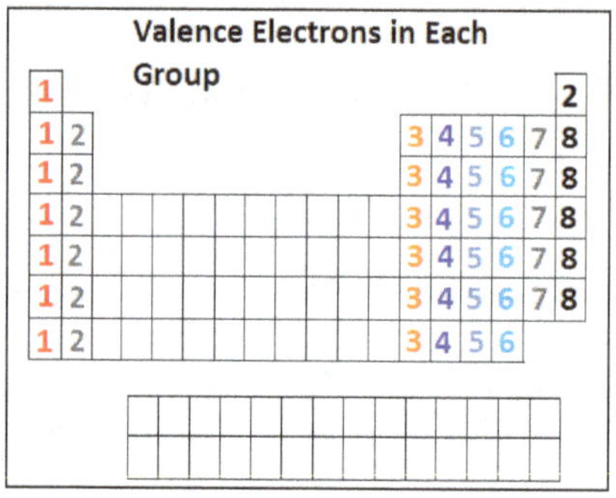

Lewis Diagrams for Molecular Compounds/Ions

To draw the lewis diagrams for molecular compounds or ions, follow these steps below (we will be using H_2O as an example to follow):

1. Count the number of valence electrons of the molecular compound or ion. Remember, if there are two or more of the same element, then you have to double or multiply by however

many atoms there are of the number of valence electrons. Follow the roman numeral group number to see the corresponding number of valence electrons there are for that element.

Valence electrons:

- Oxygen (O)--Group VIA: Therefore, there are 6 valence electrons.
- Hydrogen (H)--Group IA: Therefore, there is 1 valence electron.

There are TWO hydrogen atoms, so multiply 1 valence electron × 2 atoms.

Total: 6 + 2 = 8 valence electrons.

2. If the molecule is an ion, remember to add or subract the respective number of electrons to the total from step 1.

   For ions, if the ion has a negative charge (anion), add the corresponding number of electrons to the total number of electrons (i.e. if NO_3^- has a negative charge of 1-, then you *add* 1 extra electron to the total; 5 + 3(6)= 23 +1 = 24 total electrons). A $^-$ sign mean the molecule has an overall negative charge, so it must have this extra electron. This is because anions have a higher electron affinity (tendency to gain electrons). Most anions are composed of nonmetals, which have high electronegativity.

   If the ion has a positive charge (cation), *subtract* the corresponding number of electrons to the total number of electrons (i.e. H_3O^+ has a positive charge of 1+, so you *subtract* 1 extra electron to the total; 6 + 1(3) = 9 - 1 = 8 total electrons). A $^+$ sign means the molecule has an overall postive charge, so it must be missing one electron. Cations are positive and have weaker electron affinity. They are mostly composed of metals; their atomic radii are larger than the nonmetals. This consequently means that shielding is increased, and electrons have less tendency to be attracted to the "shielded" nucleus.

   From our example, water is a neutral molecule, therefore no electrons need to be added or subtracted from the total.

3. Write out the symbols of the elements, making sure all atoms are accounted for (i.e. H_2O, write out O and 2 H's on either side of the oxygen). Start by adding single bonds (1 pair of electrons) to all possible atoms while making sure they follow the octet rule.

4. If there are any leftover electrons, then add them to the central atom of the molecule (i.e. XeF_4 has 4 extra electrons after being distributed, so th$_4$ extra electrons are given to Xe: like so. Finally, rearrange the electron pairs into double or triple bonds if possible.

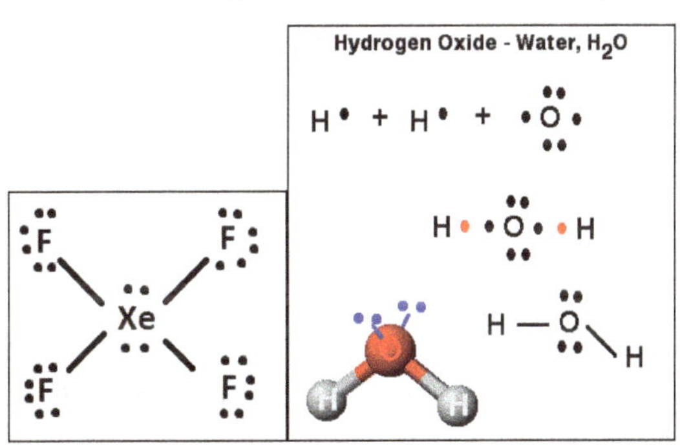

Octet Rule

Most elements follow the octet rule in chemical bonding, which means that an element should have contact to eight valence electrons in a bond or exactly fill up its valence shell. Having eight electrons total ensures that the atom is stable. This is the reason why noble gases, a valence electron shell of 8 electrons, are chemically inert; they are already stable and tend to not need the transfer of electrons when bonding with another atom in order to be stable. On the other hand, alkali metals have a valence electron shell of one electron. Since they want to complete the octet rule they often simply lose one electron. This makes them quite reactive because they can easily donate this electron to other elements. This explains the highly reactive properties of the Group IA elements.

Some elements that are exceptions to the octet rule include Aluminum(Al), Phosphorus(P), Sulfur(S), and Xenon(Xe).

Hydrogen(H) and Helium(He) follow the duet rule since their valence shell only allows two electrons. There are no exceptions to the duet rule; hydrogen and helium will always hold a maximum of two electrons.

Ionic Bonding

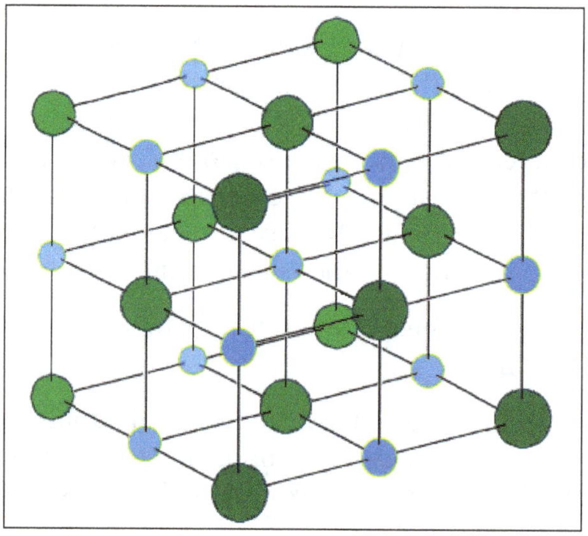

The Sodium Chloride Crystal Structure: Each atom has six nearest neighbors, with octahedral geometry. This arrangement is known as c close packed *(ccp)*. (Light blue = Na^+, Dark green = Cl^-)

Ionic bonding is when positively and negatively charged ions stick to each other through electrostatic force. These bonds are slightly weaker than covalent bonds and stronger than Van der Waals bonding or hydrogen bonding.

In ionic bonds the electronegativity of the negative ion is so much stronger than the electronegativity of the positive ion that the two ions do not share electrons. Rather, the more electronegative ion assumes full ownership of the electrons.

Perhaps the most common example of an ionically bonded substance is NaCl, or table salt. In this, the sodium (Na) atom gives up an electron to the much more electronegative chlorine (Cl) atom, and the two atoms become ions, Na^+ and Cl^-. The electrostatic bonding force between the two oppositely charged ions extends outside the local area attracting other ions to form giant crystal structures. For this reason most ionically bonded materials are solid at room temperature.

Sodium chloride forms crystals with cubic symmetry. In these, the larger chloride ions are arranged in a cubic close-packing, while the smaller sodium ions fill the octahedral gaps between them. Each ion is surrounded by six of the other kind. This same basic structure is found in many other minerals, and is known as the halite structure.

Covalent Bonding

Covalent bonding is close to the heart of organic chemistry. This is where two atoms share electrons in a bond. The goal of each atom is to fill its octet as well as have a formal charge of zero. To do this, atomic nuclei share electrons in the space between them. This sharing also allows the atoms to reach a lower energy state, which stabilizes the molecule. Most reactions in chemistry are due to molecules achieving a lower energy state. Covalent bonds are most frequently seen between atoms with similar electronegativity. In molecules that only have one type of atom, e.g. H_2 or O_2, the electronegativity of the atoms is essentially identical, so they cannot form ionic bonds. They always form covalent bonds.

Carbon is especially good at covalent bonding because its electronegativity is intermediate relative to other atoms. That means it can give as well as take electrons as needs warrant.

Covalently bonded compounds have strong internal bonds but weak attractive forces between molecules. Because of these weak attractive forces, the melting and boiling points of these compounds are much lower than compounds with ionic bonds. Therefore, such compounds are much more likely to be liquids or gases at room temperature than ionically bonded compounds.

In molecules formed from two atoms of the same element, there is no difference in the electronegativity of the bonded atoms, so the electrons in the covalent bond are shared equally, resulting in a completely non-polar covalent bond. In covalent bonds where the bonded atoms are different elements, there is a difference in electronegativities between the two atoms. The atom that is more electronegative will attract the bonding electrons more toward itself than the less electronegative atom. The difference in charge on the two atoms because of the electrons causes the covalent bond to be polar. Greater differences in electronegativity result in more polar bonds. Depending on the difference in electronegativities, the polarity of a bond can range from non-polar covalent to ionic with varying degrees of polar covalent in between. An overall imbalance in charge from one side of a molecule to the other side is called a dipole moment. Such molecules are said to be polar. For a completely symmetrical covalently bonded molecule, the overall dipole moment of the molecule is zero. Molecules with larger dipole moments are more polar. The most common polar molecule is water.

Bond Polarity and Dipole Moment

The ideas of bond polarity and dipole moment play important roles in organic chemistry.

If you look at the image of methane on the right, the single most important aspect of it in terms of bond polarity is that it is a symmetric molecule. It has 4 hydrogens, all bonded at 109.5° from the other, and all with precisely the same bond angle. Each carbon-hydrogen bond is slightly polar (hydrogen has an electronegativity of 2.1, carbon 2.5), but because of this symmetry, the polarities cancel each other out and overall, methane is a non-polar molecule.

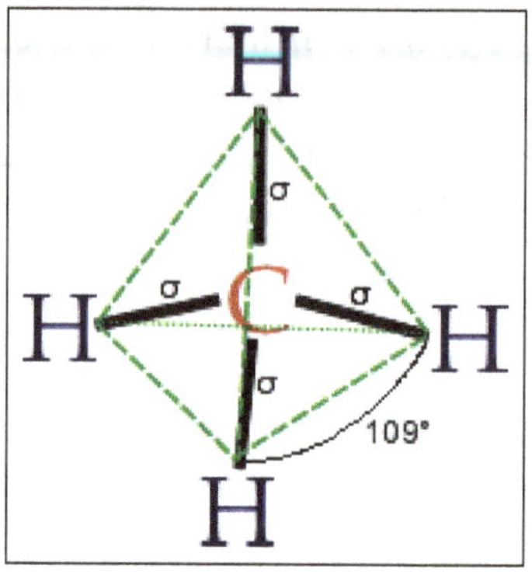

Methane

The distinction is between Bond Polarity and Molecular polarity. The total polarity of a molecule is measured as Dipole Moment. The actual calculation of dipole moment isn't really necessary so much as an understanding of what it means. Frequently, a guesstimate of dipole moment is pretty easy once you understand the concept and until you get into the more advanced organic chemistry, exact values are of little value.

Basically, the molecular polarity is, essentially, the summation of the vectors of all of the bond polarities in a molecule.

Van Der Waals Bonding

Van der Waals bonding is the collective name for three types of interactions:

1. Permanent Dipole interactions: These are the electrostatic attractive forces between two dipoles, these are responsible for fluromethane's (CH_3F) high boiling point (about -15 °C) compared to Nitrogen (about -180 °C).

2. Permanent dipole/induced dipole: These are the interactions between a permanent dipole and another molecule, causing the latter molecule's electron cloud to be distorted and thus have an induced dipole itself. These are much weaker than the permanent dipole/dipole interactions. These forces occur in permanent dipole-molecules and in mixtures of permanent dipole and dipole free molecules.

3. Instantaneous dipole/induced dipole: At any specific moment the electron cloud is not necessarily symmetrical, this instantaneous dipole then induces a dipole in another molecule and they are attracted; this is the weakest of all molecular interactions.

A Dipole is caused by an atom or molecule fragment having a higher electronegativity (this is a measure of its effective nuclear charge, and thus the attraction of the nucleus by electrons) than one to which it is attached. This means that it pulls electrons closer to it, and has a higher share of the electrons in the bond. Dipoles can cancel out by symmetry, eg: Carbon dioxide (O=C=O) is linear so there is no dipole, but the charge distribution is asymmetric causing a quadrupole moment (this acts similarly to a dipole, but is much weaker).

Organometallic Compounds and Bonding

Organometallic chemistry combines aspects of inorganic chemistry and organic chemistry, because organometallic compounds are chemical compounds containing bonds between carbon and a metal or metalloid element. Organometallic bonds are different from other bonds in that they are not either truly covalent or truly ionic, but each type of metal has individual bond character. Cuprate (copper) compounds, for example, behave quite differently than Grignard reagents (magnesium), and so beginning organic chemists should concentrate on how to use the most basic compounds mechanistically.

2

Organic Compounds

The chemical compounds which contain carbon are known as organic compounds. There are diverse types of organic compounds such as hydrocarbons, organometallic compounds, volatile organic compounds and semi-volatile organic compounds. This chapter discusses in detail these organic compounds and their properties.

An organic compound is defined as any compound whose molecules contain carbon and hydrogen (also known as "hydrocarbons") or compound that is the derivative of it. The branch of science which deals with the scientific study of structure, properties and reactions of hydrocarbons and their derivatives is known as organic chemistry.

Characteristics of Organic Compounds

The general characteristics of Organic Compounds include:

- Can be isolated as well as prepared in laboratory.
- Comprise almost 90% of all known compounds.
- Mostly built up of only three elements- carbon, hydrogen and oxygen. Other elements like halogen, nitrogen as well as phosphorous are also present but to a lesser extent.
- Possess complex structures and high molecular weights.
- Their properties are decided by certain active atom or group of atoms known as the functional group.
- They are mostly insoluble in water but soluble in organic solvents.
- They are combustible in nature.
- Chemical reactions involving organic compounds proceed at slower rates.

Characteristics due to Presence of Covalent Bonds

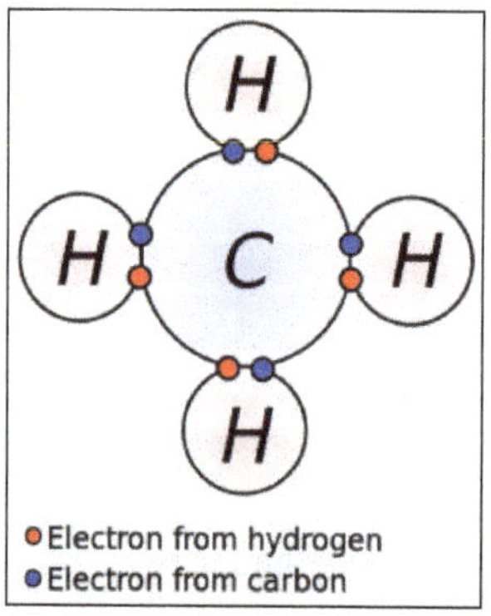

A covalent bond is a chemical *bond* that involves the sharing of electron pairs between atoms that results in a stable balance of attractive and repulsive forces between the atoms. The presence of a covalent bond renders certain characteristics to the organic compounds. These include:

- Low melting points and boiling points in comparison to the inorganic compounds.
- Organic acids and bases are less stronger and thus they have a limited dissociation in an aqueous medium.
- They exhibit the phenomenon of isomerism in which a single molecular formula represents several organic compounds differing in physical and chemical properties.
- They are volatile in nature.

General Characteristics of Members of Homologous Series

A homologous series is a group of organic chemical compounds, usually listed in order of increasing size, that have a similar structure (and hence, also similar properties) and whose structures differ only by the number of $CH_2 - CH_2$ units in the main carbon chain. They possess the following general characteristics:

- A general formula describes the members of the homologous series.
- Successive members differ from each other by CH_2CH_2.
- Physical properties change regularly with increasing number of carbon atoms.
- Members have similar chemical properties because they have same functional group.
- Members of the homologous series can be prepared using the same method.

Importance of Organic Compounds

- Organic compounds are important because all living organisms contain carbon.
- While carbohydrates, proteins and fats, the basic structures of life, are organic compounds.
- They are the basic components of many of the cycles that drive the earth. For example, the carbon cycle that includes the exchange of carbon between plants and animals in photosynthesis and cellular respiration.
- Organic compounds combine with metals to form organometallic compounds. These compounds are important industrially. They are used as catalysts, promoters, analysers as well as stabilizers.

Hydrocarbons

Hydrocarbon is any of a class of organic chemical compounds composed only of the elements carbon (C) and hydrogen (H). The carbon atoms join together to form the framework of the compound, and the hydrogen atoms attach to them in many different configurations. Hydrocarbons are the principal constituents of petroleum and natural gas. They serve as fuels and lubricants as well as raw materials for the production of plastics, fibres, rubbers, solvents, explosives, and industrial chemicals.

Many hydrocarbons occur in nature. In addition to making up fossil fuels, they are present in trees and plants, as, for example, in the form of pigments called carotenes that occur in carrots and green leaves. More than 98 percent of natural crude rubber is a hydrocarbon polymer, a chainlike molecule consisting of many units linked together. The structures and chemistry of individual hydrocarbons depend in large part on the types of chemical bonds that link together the atoms of their constituent molecules.

Nineteenth-century chemists classified hydrocarbons as either aliphatic or aromatic on the basis of their sources and properties.

Aliphatic Hydrocarbons

An aliphatic compound is a hydrocarbon compound containing carbon and hydrogen joined together in straight chains, branched trains or non-aromatic rings. Aliphatic compounds may be saturated (e.g., hexane and other alkanes) or unsaturated (e.g., hexene and other alkenes, as well as alkynes).

The simplest aliphatic hydrocarbon is methane, CH_4. In addition to hydrogen, other elements may be bound to the carbon atoms in the chain, including oxygen, nitrogen, chlorine, and sulfur. Most aliphatic hydrocarbons are flammable.

Saturated Aliphatic Hydrocarbons

A saturated hydrocarbon or saturated aliphatic hydrocarbons is a hydrocarbon in which all the carbon-carbon bonds are single bonds. A hydrocarbon is an organic compound whose only constituents are carbon and hydrogen. As the name suggests, saturated hydrocarbons are hydrocarbons in which all the carbon atoms are bonded to four other atoms and are 'saturated', implying that no carbon-carbon multiple bonds exist in these organic compounds.

Generally, the term 'saturated hydrocarbon' is used to refer to alkanes.

Alkanes

Alkanes, hydrocarbons in which all the bonds are single, have molecular formulas that satisfy the general expression C_nH_{2n+2} (where n is an integer). Carbon is sp^3 hybridized (three electron pairs are involved in bonding, forming a tetrahedral complex), and each C—C and C—H bond is a sigma (σ) bond. In order of increasing number of carbon atoms, methane (CH_4), ethane (C_2H_6), and propane (C_3H_8) are the first three members of the series.

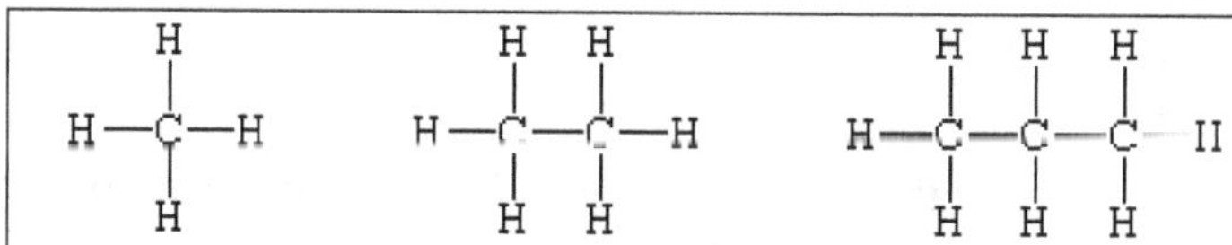

methane (CH_4) ethane (C_2H_6) propane (C_3H_8)

Methane, ethane, and propane are the only alkanes uniquely defined by their molecular formula. For C_4H_{10} two different alkanes satisfy the rules of chemical bonding (namely, that carbon has four bonds and hydrogen has one in neutral molecules). One compound, called n-butane, where the prefix n- represents normal, has its four carbon atoms bonded in a continuous chain. The other, called isobutane, has a branched chain.

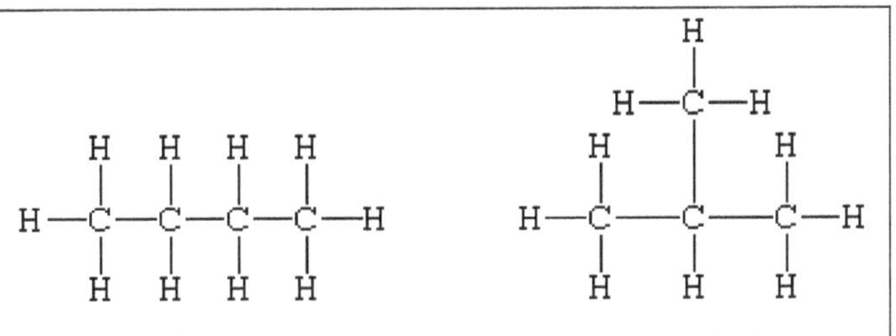

n-butane, $CH_3CH_2CH_2CH_3$ isobutane, $(CH_3)_3CH$

Different compounds that have the same molecular formula are called isomers. Isomers that differ in the order in which the atoms are connected are said to have different constitutions and are referred to as constitutional isomers. (An older name is structural isomers.) The compounds n-butane and isobutane are constitutional isomers and are the only ones possible for the formula C_4H_{10}. Because isomers are different compounds, they can have different physical and chemical properties. For example, n-butane has a higher boiling point (−0.5 °C [31.1 °F]) than isobutane (−11.7 °C [10.9 °F]).

There is no simple arithmetic relationship between the number of carbon atoms in a formula and the number of isomers. Graph theory has been used to calculate the number of constitutionally isomeric alkanes possible for values of n in C_nH_{2n+2} from 1 through 400. The number of constitutional isomers increases sharply as the number of carbon atoms increases. There is probably no upper limit to the number of carbon atoms possible in hydrocarbons. The alkane $CH_3(CH_2)_{388}CH_3$, in which 390 carbon atoms are bonded in a continuous chain, has been synthesized as an example of a so-called superlong alkane. Several thousand carbon atoms are joined together in molecules of hydrocarbon polymers such as polyethylene, polypropylene, and polystyrene.

Number of possible alkane isomers	
Molecular formula	Number of constitutional isomers
C_3H_8	1
C_4H_{10}	2
C_5H_{12}	3
C_6H_{14}	5
C_7H_{16}	9
C_8H_{18}	18
C_9H_{20}	35
$C_{10}H_{22}$	75
$C_{15}H_{32}$	4,347
$C_{20}H_{42}$	366,319
$C_{30}H_{62}$	4,111,846,763

Nomenclature

The need to give each compound a unique name requires a richer variety of terms than is available with descriptive prefixes such as n- and iso-. The naming of organic compounds is facilitated through the use of formal systems of nomenclature. Nomenclature in organic chemistry is of two types: common and systematic. Common names originate in many different ways but share the feature that there is no necessary connection between name and structure. The name that corresponds to a specific structure must simply be memorized, much like learning the name of a person. Systematic names, on the other hand, are keyed directly to molecular structure according to a generally agreed upon set of rules. The most widely used standards for organic nomenclature evolved from suggestions made by a group of chemists assembled for that purpose in Geneva in 1892 and have been revised on a regular basis by the International Union of Pure and Applied Chemistry (IUPAC). The IUPAC rules govern all classes of organic compounds but are ultimately based on alkane names. Compounds in other families are viewed as derived from alkanes by appending functional groups to, or otherwise modifying, the carbon skeleton.

The IUPAC rules assign names to unbranched alkanes according to the number of their carbon atoms. Methane, ethane, and propane are retained for CH_4, CH_3CH_3, and $CH_3CH_2CH_3$, respectively. The n-prefix is not used for unbranched alkanes in systematic IUPAC nomenclature; therefore, $CH_3CH_2CH_2CH_3$ is defined as butane, not n-butane. Beginning with five-carbon chains, the names of unbranched alkanes consist of a Latin or Greek stem corresponding to the number of carbons in the chain followed by the suffixane. A group of compounds such as the unbranched alkanes that differ from one another by successive introduction of CH_2 groups constitute a homologous series.

IUPAC Names of Unbranched Alkanes			
Alkane Formula	Name	Alkane Formula	Name
CH_4	Methane	$CH_3(CH_2)_6CH_3$	Octane
CH_3CH_3	Ethane	$CH_3(CH_2)_7CH_3$	Nonane
$CH_3CH_2CH_3$	Propane	$CH_3(CH_2)_8CH_3$	Decane
$CH_3CH_2CH_2CH_3$	Butane	$CH_3(CH_2)_{13}CH_3$	Pentadecane
$CH_3(CH_2)_3CH_3$	Pentane	$CH_3(CH_2)_{18}CH_3$	Icosane
$CH_3(CH_2)_4CH_3$	Hexane	$CH_3(CH_2)_{28}CH_3$	Triacontane
$CH_3(CH_2)_5CH_3$	Heptane	$CH_3(CH_2)_{98}CH_3$	Hectane

Alkanes with branched chains are named on the basis of the name of the longest chain of carbon atoms in the molecule, called the parent. The alkane shown has seven carbons in its longest chain and is therefore named as a derivative of heptane, the unbranched alkane that contains seven carbon atoms. The position of the CH_3 (methyl) substituent on the seven-carbon chain is specified by a number (3-), called a locant, obtained by successively numbering the carbons in the parent chain starting at the end nearer the branch. The compound is therefore called 3-methylheptane.

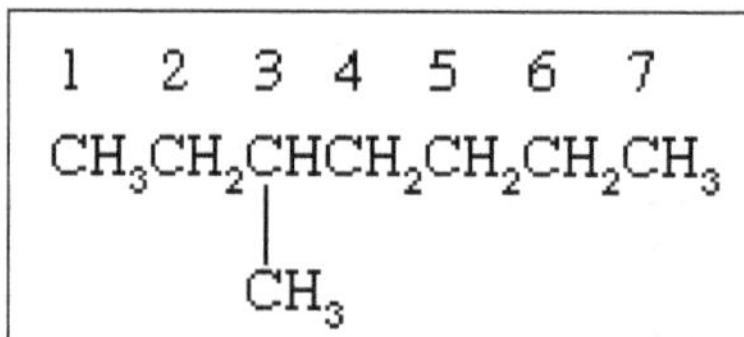

3-methylheptane

When there are two or more identical substituents, replicating prefixes (di-, tri-, tetra-, etc.) are used, along with a separate locant for each substituent. Different substituents, such as ethyl ($-CH_2CH_3$) and methyl ($-CH_3$) groups, are cited in alphabetical order. Replicating prefixes are ignored when alphabetizing. In alkanes, numbering begins at the end nearest the substituent that appears first on the chain so that the carbon to which it is attached has as low a number as possible.

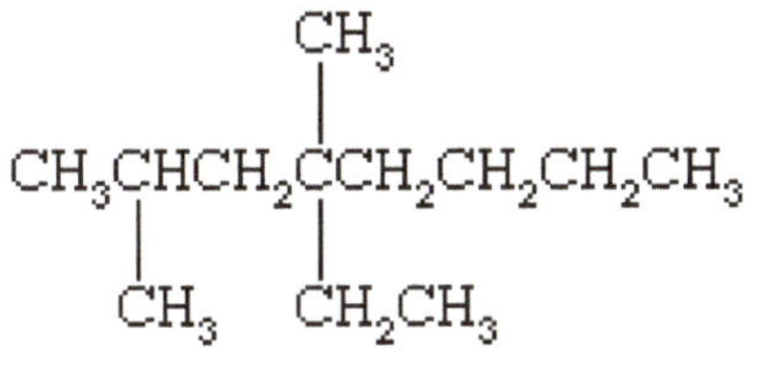

4-ethyl-2,4-dimethyloctane

Methyl and ethyl are examples of alkyl groups. An alkyl group is derived from an alkane by deleting one of its hydrogens, thereby leaving a potential point of attachment. Methyl is the only alkyl group derivable from methane and ethyl the only one from ethane. There are two C_3H_7 and four C_4H_9 alkyl groups. The IUPAC rules for naming alkanes and alkyl groups cover even very complex structures and are regularly updated. They are unambiguous in the sense that, although a single compound may have more than one correct IUPAC name, there is no possibility that two different compounds will have the same name.

Three-dimensional Structures

Most organic molecules, including all alkanes, are not planar but are instead characterized by three-dimensional structures. Methane, for example, has the shape of a regular tetrahedron with carbon at the centre and a hydrogen atom at each corner. Each H–C–H angle in methane is 109.5°, and each C–H bond distance is 1.09 angstroms (Å; 1Å = 1×10^{-10} metre). Higher alkanes such as butane have bonds that are tetrahedrally disposed on each carbon except that the resulting C–C–C and H–C–H angles are slightly larger and smaller, respectively, than the ideal value of 109.5° characteristic of a perfectly symmetrical tetrahedron. Carbon-carbon bond distances in alkanes are normally close to 1.53 angstroms.

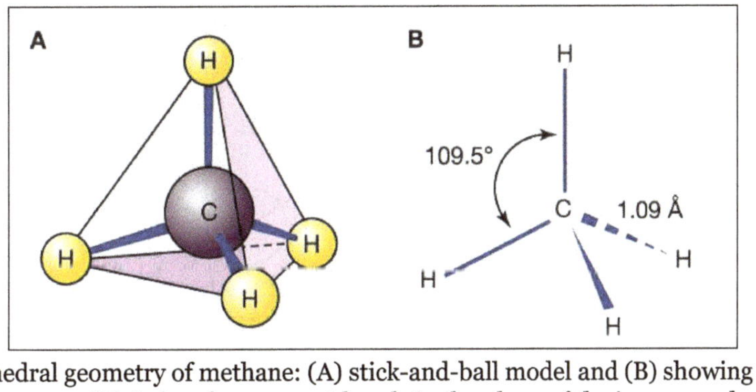

Tetrahedral geometry of methane: (A) stick-and-ball model and (B) showing bond angles and distances. (Plain bonds represent bonds in the plane of the image; wedge and dashed bonds represent those directed toward and away from the viewer, respectively.)

An important aspect of the three-dimensional shape of alkanes and other organic molecules is their conformations, the nonidentical arrangements of atoms that are generated by rotation about single bonds. Of the infinite number of conformations possible for ethane—which are related by tiny increments of rotation of one CH_3 group with respect to the other—the eclipsed conformation is the least stable, and the staggered conformation is the most stable. The eclipsed conformation is said to suffer torsional strain because of repulsive forces between electron pairs in the C–H bonds of adjacent carbons. These repulsive forces are minimized in the staggered conformation since all C–H bonds are as far from one another as possible. Although rotation about the C–C bond of ethane is exceedingly rapid (millions of times per second at room temperature), at any instant most of the molecules exist in the staggered conformation.

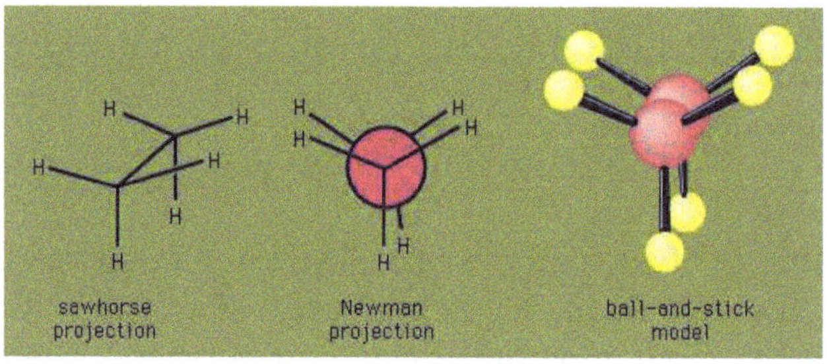

Eclipsed conformation is the least stable of all ethane conformations because the repulsive forces between electron pairs in the C–H bonds of adjacent carbons are maximized.

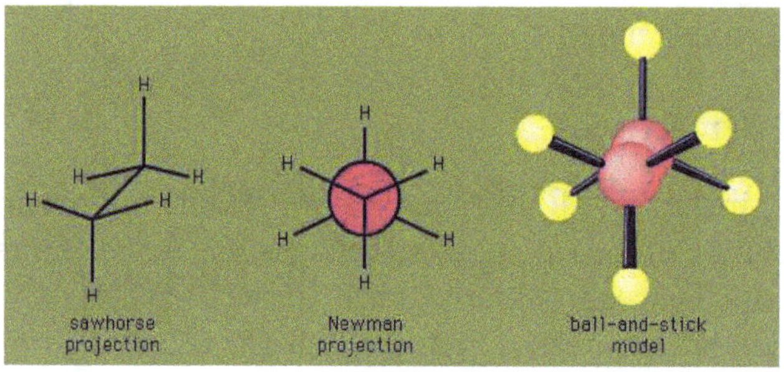

Staggered conformation is the most stable of all ethane conformations because the repulsive forces between electron pairs in the C–H bonds of adjacent carbons are minimized.

For butane, two different staggered conformations, called anti and gauche, are possible. Methyl is a larger substituent than hydrogen, and the greater separation between methyl groups in the anti conformation makes it slightly more stable than the gauche.

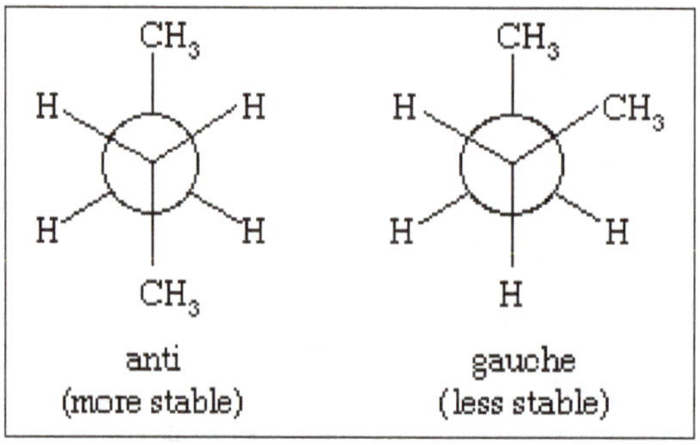

The three-dimensional structures of higher alkanes are governed by the tetrahedral disposition of the four bonds to each carbon atom, by the preference for staggered conformations, and by the greater stability of anti C–C–C–C arrangements over gauche.

Cycloalkanes

Countless organic compounds are known in which a sequence of carbon atoms, rather than being connected in a chain, closes to form a ring. Saturated hydrocarbons that contain one ring are referred to as cycloalkanes. With a general formula of C_nH_{2n} (n is an integer greater than 2), they have two fewer hydrogen atoms than an alkane with the same number of carbon atoms. Cyclopropane (C_3H_6) is the smallest cycloalkane, whereas cyclohexane (C_6H_{12}) is the most studied, best understood, and most important. It is customary to represent cycloalkane rings as polygons, with the understanding that each corner corresponds to a carbon atom to which is attached the requisite number of hydrogen atoms to bring its total number of bonds to four.

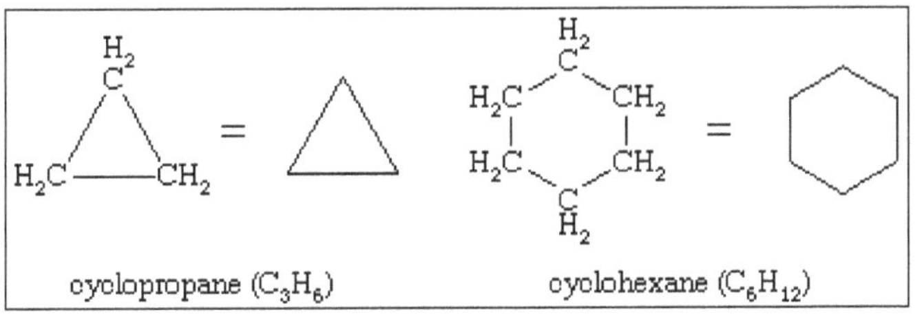

In naming cycloalkanes, alkyl groups attached to the ring are indicated explicitly and listed in alphabetical order, and the ring is numbered so as to give the lowest locant to the first-appearing substituent. If two different directions yield equivalent locants, the direction is chosen that gives the lower number to the substituent appearing first in the name.

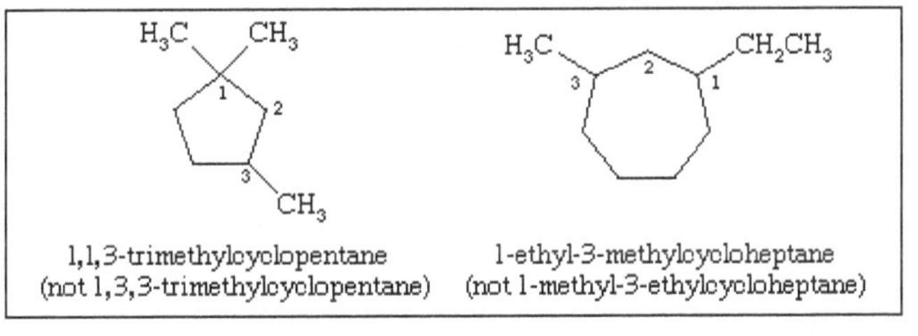

The three carbon atoms of cyclopropane define the corners of an equilateral triangle, a geometry that requires the C–C–C angles to be 60°. This 60° angle is much smaller than the normal tetrahedral bond angle of 109.5° and imposes considerable strain (called angle strain) on cyclopropane. Cyclopropane is further destabilized by the torsional strain that results from having three eclipsed C–H bonds above the plane of the ring and three below.

Cyclopropane is the only cycloalkane that is planar. Cyclobutane (C_4H_8) and higher cycloalkanes adopt nonplanar conformations in order to minimize the eclipsing of bonds on adjacent atoms. The angle strain in cyclobutane is less than in cyclopropane, whereas cyclopentane and higher cycloalkanes are virtually free of angle strain. With the exception of cyclopropane, all cycloalkanes undergo rapid internal motion involving interconversion of nonplanar "puckered" conformations.

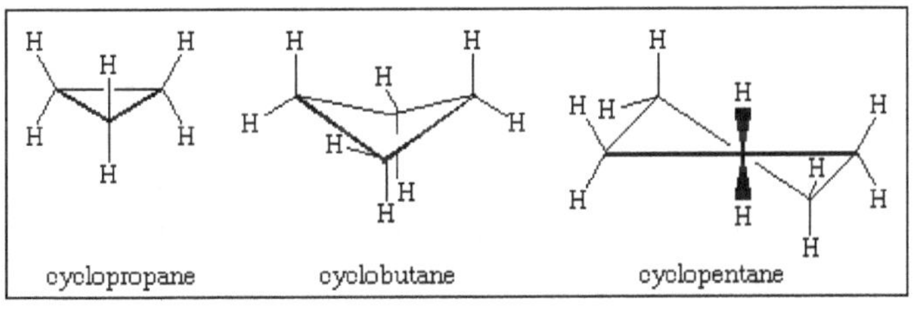

Many of the most important principles of conformational analysis have been developed by examining cyclohexane. Three conformations of cyclohexane, designated as chair, boat, and skew (or twist), are essentially free of angle strain. Of these three the chair is the most stable, mainly because it has a staggered arrangement of all its bonds. The boat and skew conformations lack perfect staggering of bonds and are destabilized by torsional strain. The boat conformation is further destabilized by the mutual crowding of hydrogen atoms at carbons one and four. The shape of the boat brings its two "flagpole" hydrogen atoms to within 1.80 angstroms of each other, far closer than the 2.20 angstrom distance at which repulsive forces between hydrogen atoms become significant. At room temperature, 999 of every 1,000 cyclohexane molecules exist in the chair form (the other being skew).

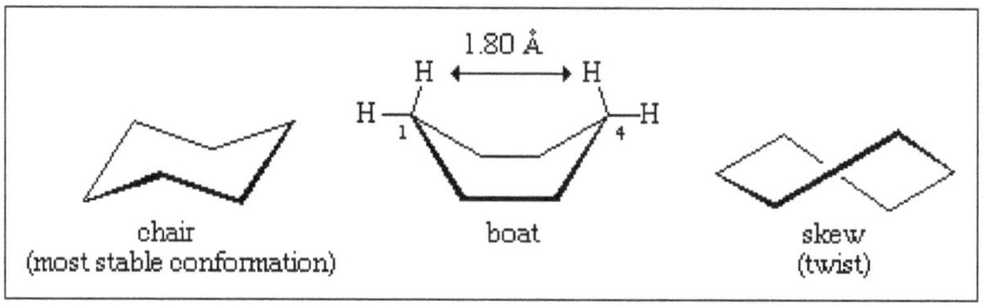

There are two orientations of carbon-hydrogen bonds in the chair conformation of cyclohexane. Six bonds are parallel to a vertical axis passing through the centre of the ring and are called axial (a) bonds. The directions of these six axial bonds alternate up and down from one carbon to the next around the ring; thus, the axial hydrogens at carbons one, three, and five lie on one side of the ring and those at carbons two, four, and six on the other. The remaining six bonds are referred to as equatorial (e) because they lie in a region corresponding to the approximate "equator" of the molecule. The shortest distances between nonbonded atoms are those involving axial hydrogens on the same side of the molecule.

A rapid process of chair-chair interconversion (called ring-flipping) interconverts the six axial and six equatorial hydrogen atoms in cyclohexane. Chair-chair interconversion is a complicated process brought about by successive conformational changes within the molecule. It is different from simple whole-molecule motions, such as spinning and tumbling, and because it is a conformational change only, it does not require any bonds to be broken.

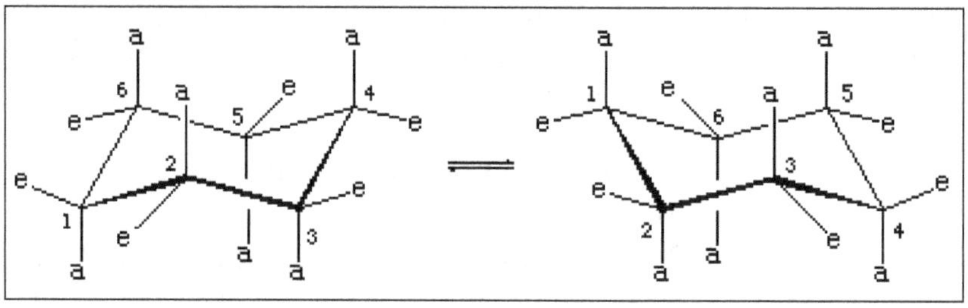

Chair-chair interconversion is especially important in substituted derivatives of cyclohexane. Any substituent is more stable when it occupies an equatorial rather than an axial site on the ring, since equatorial substituents are less crowded than axial ones. In methylcyclohexane, the chair conformation in which the large methyl group is equatorial is the most stable and, therefore, the most populated of all possible conformations. At any instant, almost all the methylcyclohexane molecules in a given sample exist in chair conformations, and about 95 percent of these have the methyl group in an equatorial orientation.

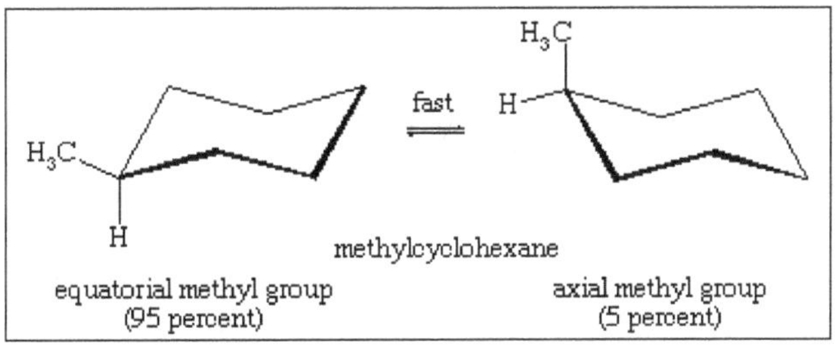

The highly branched tert-butyl group $(CH_3)_3C-$ (tert-butyl) is even more spatially demanding than the methyl group, and more than 99.99 percent of tert-butylcyclohexane molecules adopt chair conformations in which the $(CH_3)_3C-$ group is equatorial.

Conformational analysis of six-membered rings, especially the greater stability of chair conformations with equatorial substituents, not only is important in the area of hydrocarbons but also is essential to an understanding of the properties of biologically important molecules, especially steroids and carbohydrates. Odd Hassel of Norway and Derek H.R. Barton of England shared the Nobel Prize for Chemistry in 1969 for their important discoveries in this area. Hassel's studies dealt with structure, while Barton showed how conformational effects influence chemical reactivity.

The most stable structures of cycloalkanes and compounds based on them have been determined by a number of experimental techniques, including X-ray diffraction and electron diffraction analyses and infrared, nuclear magnetic resonance, and microwave spectroscopies. These experimental techniques have been joined by advances in computational methods such as molecular mechanics, whereby the total strain energies of various conformations are calculated and compared (chemical bonding: Computational approaches to molecular structure). The structure with the lowest total energy is the most stable and corresponds to the best combination of bond distances, bond angles, and conformation. One benefit of such calculations is that unstable conformations, which are difficult to study experimentally, can be examined. The quality of molecular mechanics calculations is such that it is claimed that many structural features of hydrocarbons can be computed more accurately than they can be measured.

The conformations of rings with 7–12 carbons have been special targets for study by molecular mechanics. Unlike cyclohexane, in which one conformation (the chair) is much more stable than any other, cycloalkanes with 7–12 carbons are generally populated by several conformations of similar energy. Rings with more than 12 carbons are sufficiently flexible to adopt conformations that are essentially strain-free.

Polycyclic hydrocarbons are hydrocarbons that contain more than one ring. They are classified as bicyclic, tricyclic, tetracyclic, and so forth, according to the number of formal bond disconnections necessary to produce a noncyclic carbon chain. Examples include trans-decalin and adamantane—both of which are present in small amounts in petroleum—and cubane, a compound synthesized for the purpose of studying the effects of strain on chemical reactivity.

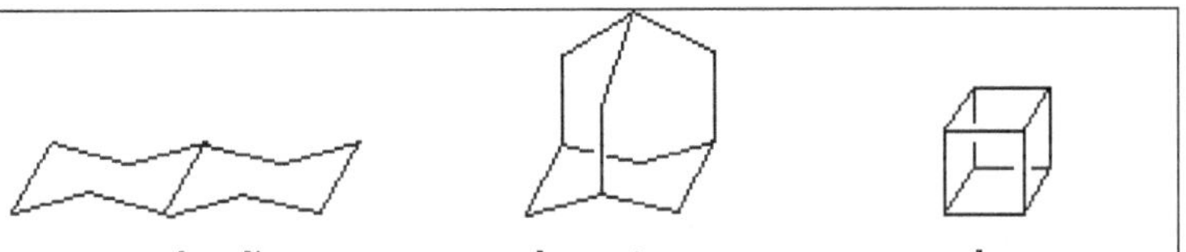

trans-decalin ($C_{10}H_{18}$; bicyclic) adamantane ($C_{10}H_{16}$; tricyclic) cubane (C_8H_8; pentacyclic)

Stereoisomerism

Certain substituted derivatives of cycloalkanes exhibit a type of isomerism called stereoisomerism in which two substances have the same molecular formula and the same constitution but differ in the arrangement of their atoms in space. Methyl groups in 1,2-dimethylcyclopropane, for example, may be on the same (cis) or opposite (trans) sides of the plane defined by the ring. The resulting two substances are different compounds, each having its own properties such as boiling point (abbreviated bp here):

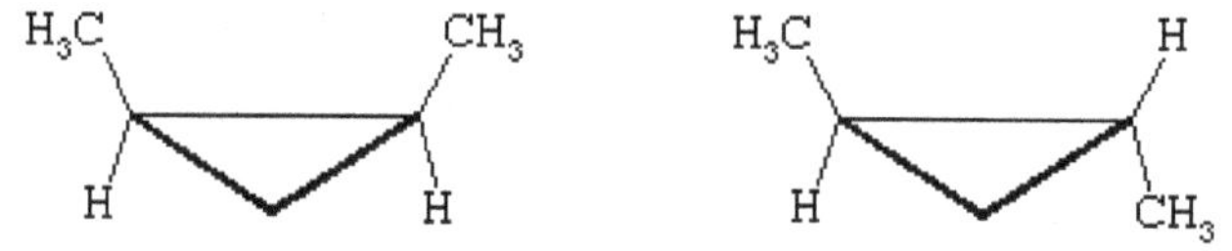

cis-1,2-dimethylcyclopropane (bp 37° C) *trans*-1,2-dimethylcyclopropane (bp 29° C)

Cis-transisomers belong to a class of stereoisomers known as diastereomers and are often referred to as geometric isomers, although this is an obsolete term. Cis-trans stereoisomers normally cannot be interconverted at room temperature, because to do so requires the breaking and reforming of chemical bonds.

Physical Properties

Alkanes and cycloalkanes are nonpolar substances. Attractive forces between alkane molecules are dictated by London forces (or dispersion forces, arising from electron fluctuations in molecules) and are weak. Thus, alkanes have relatively low boiling points compared with polar molecules of comparable molecular weight. The boiling points of alkanes increase with increasing number of carbons. This is because the intermolecular attractive forces, although individually weak, become cumulatively more significant as the number of atoms and electrons in the molecule increases.

Physical Properties of Unbranched Alkanes			
Name	Formula	Boiling Point (°C)	Melting Point (°C)
Methane	CH_4	−164	−182.5
Ethane	CH_3CH_3	−88.6	−183.3
Propane	$CH_3CH_2CH_3$	−42	−189.7
Butane	$CH_3(CH_2)_2CH_3$	−0.5	−138.35
Pentane	$CH_3(CH_2)_3CH_3$	+36.1	−129.7
Hexane	$CH_3(CH_2)_4CH_3$	+68.9	−95.0
Heptane	$CH_3(CH_2)_5CH_3$	+98.4	−90.6
Octane	$CH_3(CH_2)_6CH_3$	+125.6	−56.8
Nonane	$CH_3(CH_2)_7CH_3$	+150.8	−51.0
Decane	$CH_3(CH_2)_8CH_3$	+174.1	−29.7
Pentadecane	$CH_3(CH_2)_{13}CH_3$	+270	+10
Octadecane	$CH_3(CH_2)_{16}CH_3$	+316.1	+28.2
Icosane	$CH_3(CH_2)_{18}CH_3$	+343	+36.8

Triacontane	$CH_3(CH_2)_{28}CH_3$	+449.7	+65.8
Tetracontane	$CH_3(CH_2)_{38}CH_3$	—	+81
Pentacontane	$CH_3(CH_2)_{48}CH_3$	—	+92

For a given number of carbon atoms, an unbranched alkane has a higher boiling point than any of its branched-chain isomers. This effect is evident upon comparing the boiling points (bp) of selected C_8H_{18} isomers. An unbranched alkane has a more extended shape, thereby increasing the number of intermolecular attractive forces that must be broken in order to go from the liquid state to the gaseous state. On the other hand, the relatively compact ellipsoidal shape of 2,2,3,3-tetramethylbutane permits it to pack into a crystal lattice more effectively than octane and so raises its melting point (mp).

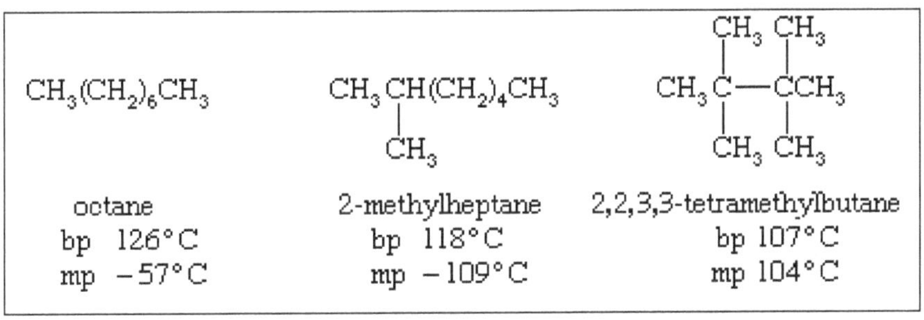

In general, solid alkanes do not often have high melting points. Unbranched alkanes tend toward a maximum in that the melting point of $CH_3(CH_2)_{98}CH_3$ (115 °C [239 °F]) is not much different from that of $CH_3(CH_2)_{148}CH_3$ (123 °C [253 °F]).

The viscosity of liquid alkanes increases with the number of carbons. Increased intermolecular attractive forces, as well as an increase in the extent to which nearby molecules become entangled when they have an extended shape, cause unbranched alkanes to be more viscous than their branched-chain isomers.

The densities of liquid hydrocarbons are all less than that of water, which is quite polar and possesses strong intermolecular attractive forces. All hydrocarbons are insoluble in water and, being less dense than water, float on its surface. Hydrocarbons are, however, usually soluble in one another as well as in organic solvents such as diethyl ether ($CH_3CH_2OCH_2CH_3$).

Sources and Occurrence

The most abundant sources of alkanes are natural gas and petroleum deposits, formed over a period of millions of years by the decay of organic matter in the absence of oxygen. Natural gas contains 60–80 percent methane, 5–9 percent ethane, 3–18 percent propane, and 2–14 percent higher hydrocarbons. Petroleum is a complex liquid mixture of hundreds of substances—including 150 or more hydrocarbons, approximately half of which are saturated.

Approximately two billion tons of methane are produced annually by the bacteria that live in termites and in the digestive systems of plant-eating animals. Smaller quantities of alkanes also can be found in a variety of natural materials. The so-called aggregation pheromone whereby Blaberus craniifer cockroaches attract others of the same species is a 1:1 mixture of the volatile but relatively high-boiling liquid alkanes undecane, $CH_3(CH_2)_9CH_3$, and tetradecane, $CH_3(CH_2)_{12}CH_3$. Hentriacontane,

$CH_3(CH_2)_{29}CH_3$, is a solid alkane present to the extent of 8–9 percent in beeswax, where its stability and impermeability to water contribute to the role it plays as a structural component.

With the exception of the alkanes that are readily available from petroleum, alkanes are synthesized in the laboratory and in industry by the hydrogenation of alkenes. Only a few methods are available in which a carbon-carbon bond-forming operation gives an alkane directly, and these tend to be suitable only for syntheses carried out on a small scale.

Chemical Reactions

As is true for all hydrocarbons, alkanes burn in air to produce carbon dioxide (CO_2) and water (H_2O) and release heat. The combustion of 2,2,4-trimethylpentane is expressed by the following chemical equation:

$(CH_3)_3CCH_2CH(CH_3)_2$ +	$^{25}/_2\,O_2$	$\longrightarrow$	$8CO_2$ +	$9H_2O$
2,2,4-trimethylpentane	oxygen		carbon dioxide	water

The fact that all hydrocarbon combustions are exothermic is responsible for their widespread use as fuels. Grades of gasoline are rated by comparing their tendency toward preignition or knocking to reference blends of heptane and 2,2,4-trimethylpentane and assigning octane numbers. Pure heptane (assigned an octane number of 0) has poor ignition characteristics, whereas 2,2,4-trimethylpentane (assigned an octane number of 100) resists knocking even in high-compression engines.

As a class, alkanes are relatively unreactive substances and undergo only a few reactions. An industrial process known as isomerization employs an aluminum chloride ($AlCl_3$) catalyst to convert unbranched alkanes to their branched-chain isomers. In one such application, butane is isomerized to 2-methylpropane for use as a starting material in the preparation of 2,2,4-trimethylpentane (isooctane), which is a component of high-octane gasoline.

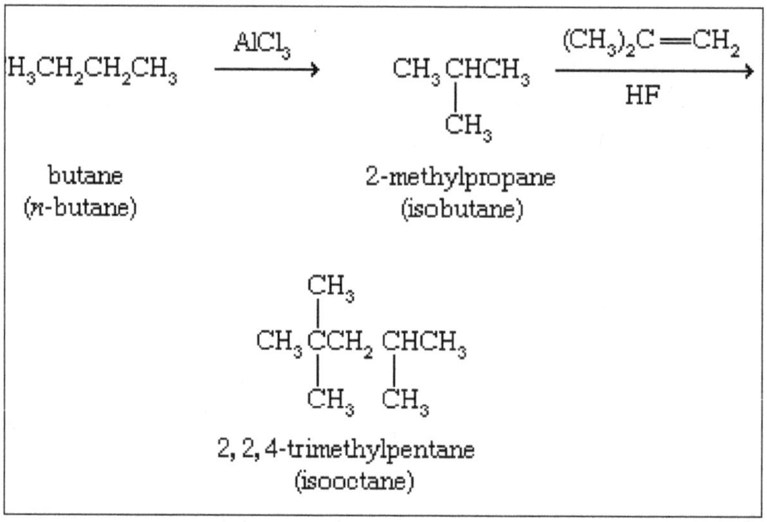

The halogens chlorine (Cl_2) and bromine (Br_2) react with alkanes and cycloalkanes by replacing one or more hydrogens with a halogen. Although the reactions are exothermic, a source of energy such as ultraviolet light or high temperature is required to initiate the reaction, as, for example, in the chlorination of cyclobutane.

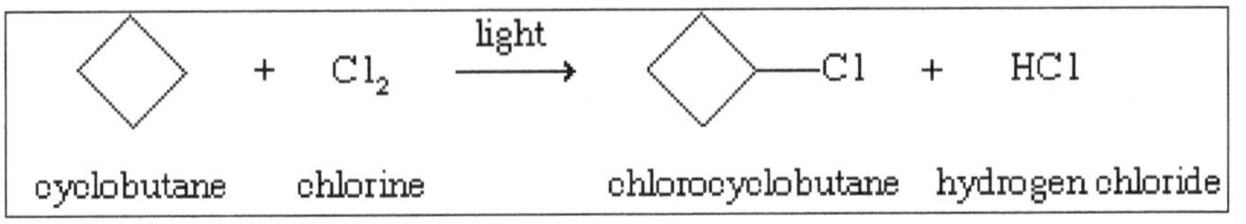

The chlorinated derivatives of methane (CH_3Cl, CH_2Cl_2, $CHCl_3$, and CCl_4) are useful industrially and are prepared by various methods, including the reaction of methane with chlorine at temperatures on the order of 450 °C (840 °F).

The most important industrial organic chemical reaction in terms of its scale and economic impact is the dehydrogenation of ethane (obtained from natural gas) to form ethylene and hydrogen (see below Alkenes and alkynes: Natural occurrence and Synthesis). The hydrogen produced is employed in the Haber-Bosch process for the preparation of ammonia from nitrogen.

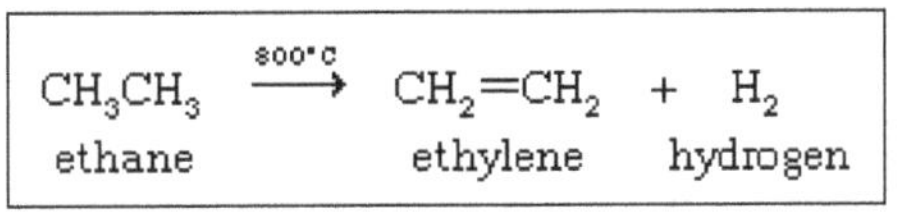

The higher alkanes present in petroleum also yield ethylene under similar conditions by reactions that involve both dehydrogenation and the breaking of carbon-carbon bonds. The conversion of high-molecular-weight alkanes to lower ones is called cracking.

Alkenes and Alkynes

Alkenes (also called olefins) and alkynes (also called acetylenes) belong to the class of unsaturated aliphatic hydrocarbons. Alkenes are hydrocarbons that contain a carbon-carbon double bond, whereas alkynes have a carbon-carbon triple bond. Alkenes are characterized by the general molecular formula C_nH_{2n}, alkynes by C_nH_{2n-2}. Ethene (C_2H_4) is the simplest alkene and ethyne (C_2H_2) the simplest alkyne.

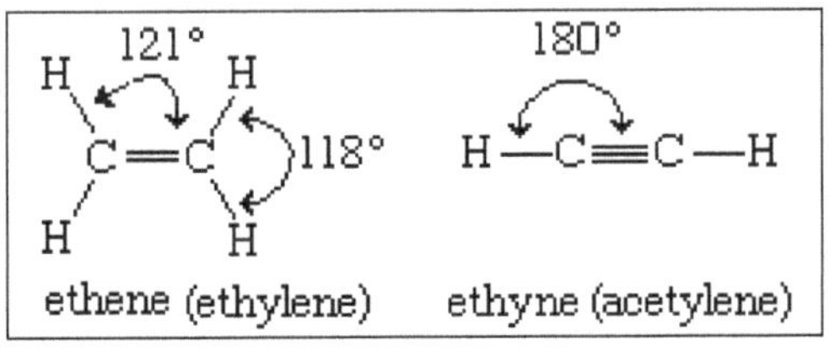

Ethylene is a planar molecule with a carbon-carbon double bond length (1.34 angstroms) that is significantly shorter than the corresponding single bond length (1.53 angstroms) in ethane. Acetylene has a linear H–C≡C–H geometry, and its carbon-carbon bond distance (1.20 angstroms) is even shorter than that of ethylene.

Bonding in Alkenes and Alkynes

The generally accepted bonding model for alkenes views the double bond as being composed of a σ (sigma) component and a π (pi) component. In the case of ethylene, each carbon is sp^2 hybridized, and each is bonded to two hydrogens and the other carbon by σ bonds. Additionally, each carbon has a half-filled p orbital, the axis of which is perpendicular to the plane of the σ bonds. Side-by-side overlap of these two p orbitals generates a π bond. The pair of electrons in the π bond are equally likely to be found in the regions of space immediately above and below the plane defined by the atoms. Most of the important reactions of alkenes involve the electrons in the π component of the double bond because these are the electrons that are farthest from the positively charged nuclei and therefore the most weakly held.

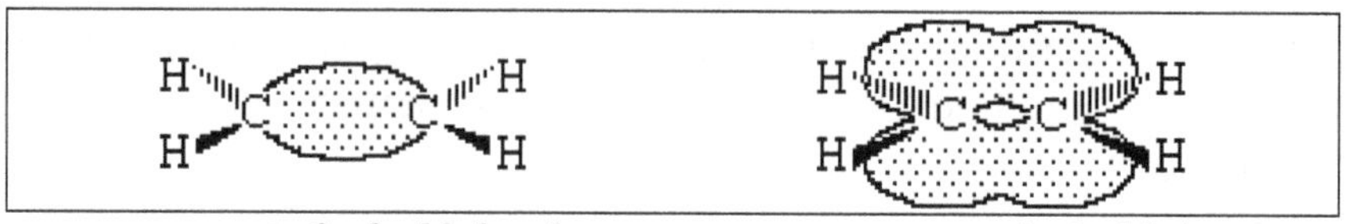

σ component of a double bond π component of a double bond

The triple bond of an alkyne consists of one σ and two π components linking two sp hybridized carbons. In the case of acetylene, the molecule itself is linear with σ bonds between the two carbons and to each hydrogen. Each carbon has two p orbitals, the axes of which are perpendicular to each other. Overlap of two p orbitals, suitably aligned and on adjacent carbons, gives two π bonds.

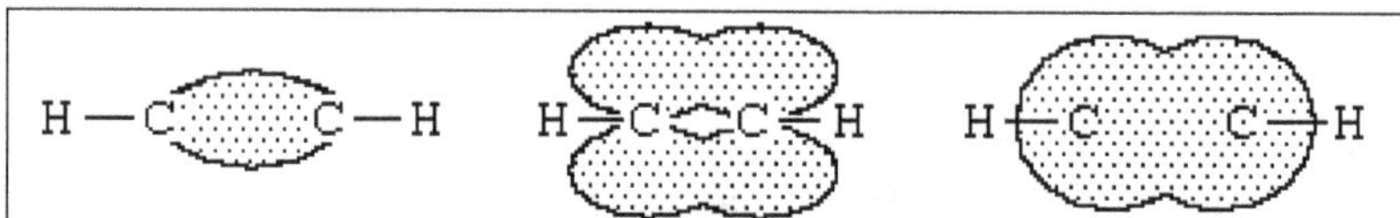

σ component of a triple bond π component of a triple bond analogous to the π bond of ethylene   "top" half of a second π component of a triple bond

Nomenclature of Alkenes and Alkynes

Ethylene and acetylene are synonyms in the IUPAC nomenclature system for ethene and ethyne, respectively. Higher alkenes and alkynes are named by counting the number of carbons in the longest continuous chain that includes the double or triple bond and appending an -ene (alkene) or -yne (alkyne) suffix to the stem name of the unbranched alkane having that number of carbons. The chain is numbered in the direction that gives the lowest number to the first multiply bonded carbon, and adding it as a prefix to the name. Once the chain is numbered with respect to the multiple bond, substituents attached to the parent chain are listed in alphabetical order and their positions identified by number.

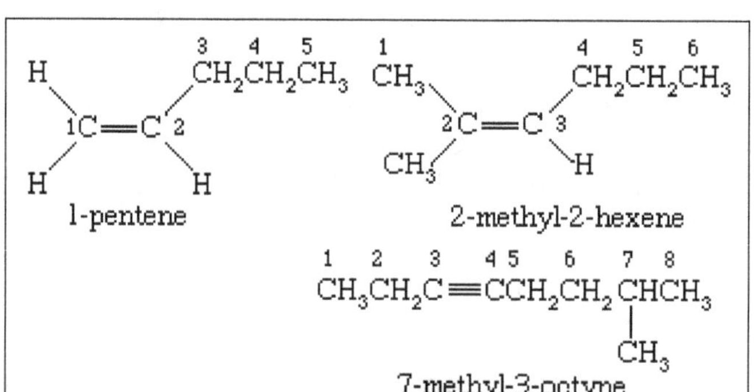

Compounds that contain two double bonds are classified as dienes, those with three as trienes, and so forth. Dienes are named by replacing the -ane suffix of the corresponding alkane by -adiene and identifying the positions of the double bonds by numerical locants. Dienes are classified

as cumulated, conjugated, or isolated according to whether the double bonds constitute a C=C=C unit, a C=C–C=C unit, or a C=C–$(CXY)_n$–C=C unit, respectively.

$CH_3CH{=}C{=}CHCH_3$ 2,3-pentadiene (cumulated)

$CH_2{=}CHCH{=}CHCH_3$ 1,3-pentadiene (conjugated)

$CH_2{=}CHCH_2CH{=}CH_2$ 1,4-pentadiene (isolated)

Double bonds can be incorporated into rings of all sizes, resulting in cycloalkenes. In naming substituted derivatives of cycloalkenes, numbering begins at and continues through the double bond.

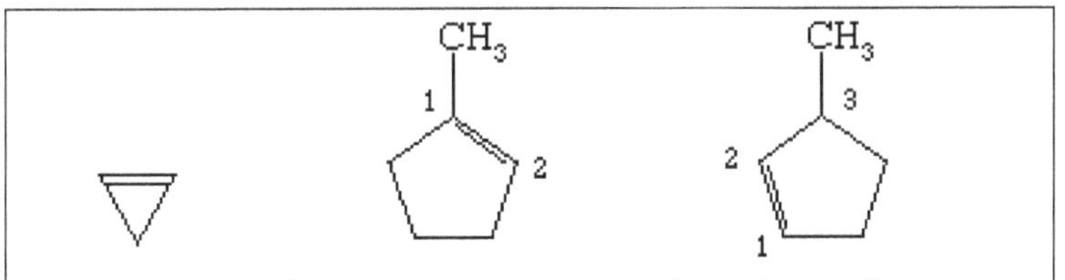

cyclopropene 1-methylcyclopentene 3-methylcyclopentene

Unlike rotation about carbon-carbon single bonds, which is exceedingly rapid, rotation about carbon-carbon double bonds does not occur under normal circumstances. Stereoisomerism is therefore possible in those alkenes in which neither carbon atom bears two identical substituents. In most cases, the names of stereoisomeric alkenes are distinguished by cis-trans notation. (An alternative method, based on the Cahn-Ingold-Prelog system and using E and Z prefixes, is also used). Cycloalkenes in which the ring has eight or more carbons are capable of existing as cis or trans stereoisomers. trans-Cycloalkenes are too unstable to isolate when the ring has seven or fewer carbons.

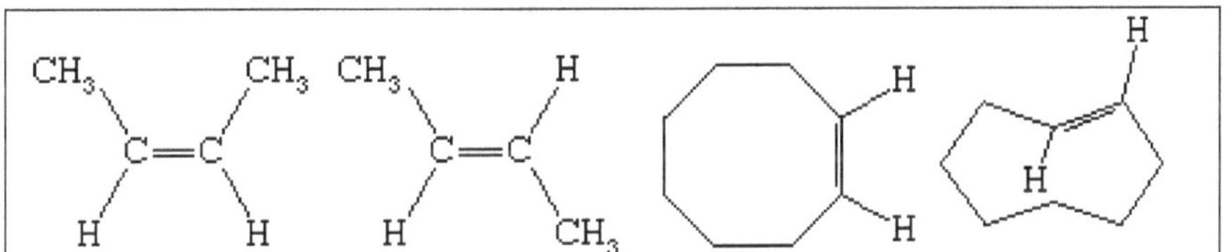

cis-2-butene *trans*-2-butene *cis*-cyclooctene *trans*-cyclooctene

Because the C–C≡C–C unit of an alkyne is linear, cycloalkynes are possible only when the number of carbon atoms in the ring is large enough to confer the flexibility necessary to accommodate this geometry. Cyclooctyne (C_8H_{12}) is the smallest cycloalkyne capable of being isolated and stored as a stable compound.

Natural Occurrence

Ethylene is formed in small amounts as a plant hormone. The biosynthesis of ethylene involves an enzyme-catalyzed decomposition of a novel amino acid, and, once formed, ethylene stimulates the ripening of fruits.

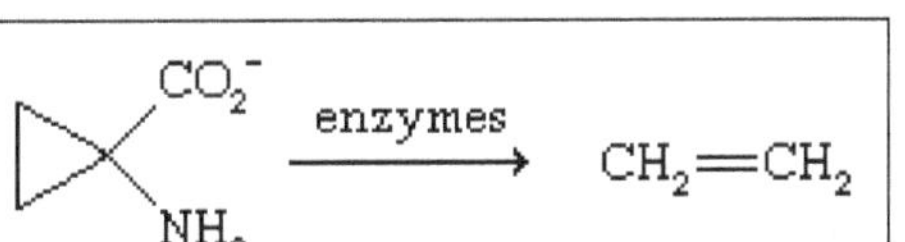

Alkenes are abundant in the essential oils of trees and other plants. (Essential oils are responsible for the characteristic odour, or "essence," of the plant from which they are obtained.) Myrcene and limonene, for example, are alkenes found in bayberry and lime oil, respectively. Oil of turpentine, obtained by distilling the exudate from pine trees, is a mixture of hydrocarbons rich in α-pinene. α-Pinene is used as a paint thinner as well as a starting material for the preparation of synthetic camphor, drugs, and other chemicals.

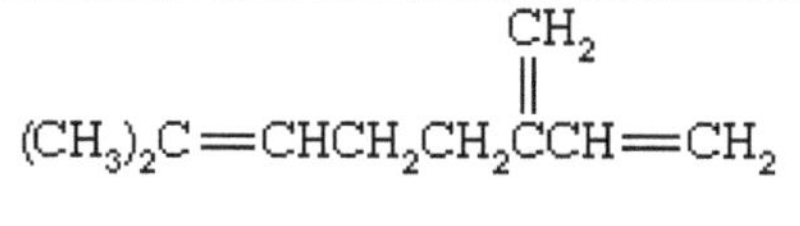

myrcene

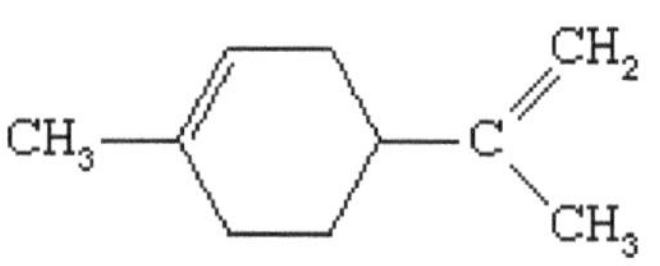

limonene

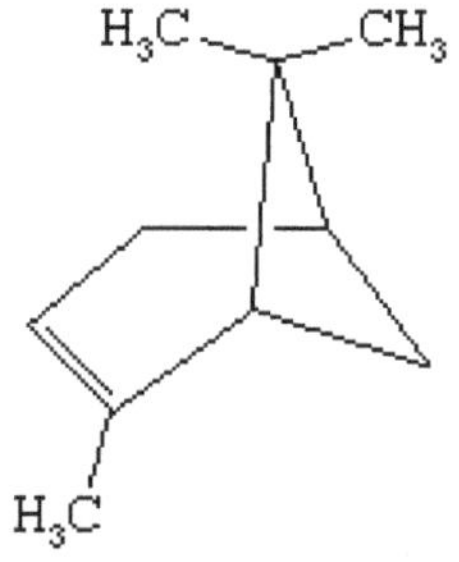

α-pinene

Other naturally occurring hydrocarbons with double bonds include plant pigments such as lycopene, which is responsible for the red colour of ripe tomatoes and watermelon. Lycopene is a polyene (meaning many double bonds) that belongs to a family of 40-carbon hydrocarbons known as carotenes.

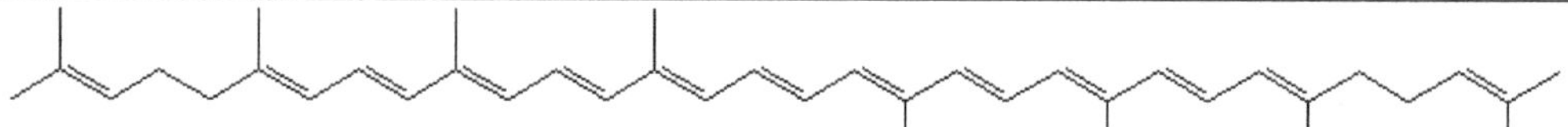

lycopene

The sequence of alternating single and double bonds in lycopene is an example of a conjugated system. The degree of conjugation affects the light-absorption properties of unsaturated compounds. Simple alkenes absorb ultraviolet light and appear colourless. The wavelength of the light absorbed by unsaturated compounds becomes longer as the number of double bonds in conjugation with one another increases, with the result that polyenes containing regions of extended conjugation absorb visible light and appear yellow to red.

The hydrocarbon fraction of natural rubber (roughly 98 percent) is made up of a collection of polymer molecules, each of which contains approximately 20,000 C_5H_8 structural units joined together in a regular repeating pattern.

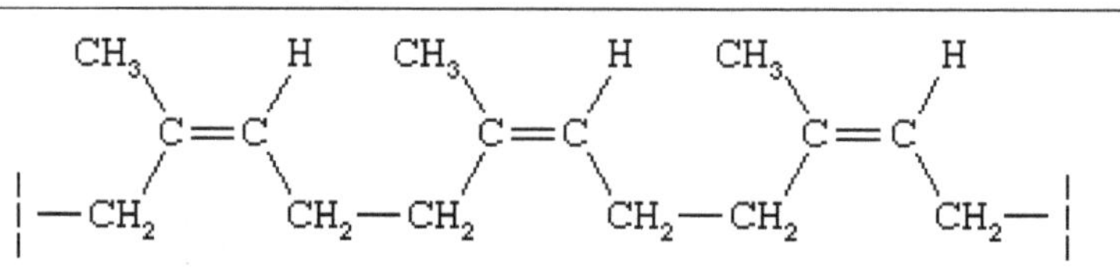

C_5H_8 units, three of which are shown, are linked together in natural rubber.

Natural products that contain carbon-carbon triple bonds, while numerous in plants and fungi, are far less abundant than those that contain double bonds and are much less frequently encountered.

Synthesis

The lower alkenes (through four-carbon alkenes) are produced commercially by cracking and dehydrogenation of the hydrocarbons present in natural gas and petroleum. The annual global production of ethylene averages around 75 million metric tons. Analogous processes yield approximately 2 million metric tons per year of 1,3-butadiene (CH_2=CHCH=CH_2). Approximately one-half of the ethylene is used to prepare polyethylene. Most of the remainder is utilized to make ethylene oxide (for the manufacture of ethylene glycol antifreeze and other products), vinyl chloride (for polymerization to polyvinyl chloride), and styrene (for polymerization to polystyrene). The principal application of propylene is in the preparation of polypropylene. 1,3-Butadiene is a starting material in the manufacture of synthetic rubber.

Higher alkenes and cycloalkenes are normally prepared by reactions in which a double bond is introduced into a saturated precursor by elimination (i.e., a reaction in which atoms or ions are lost from a molecule).

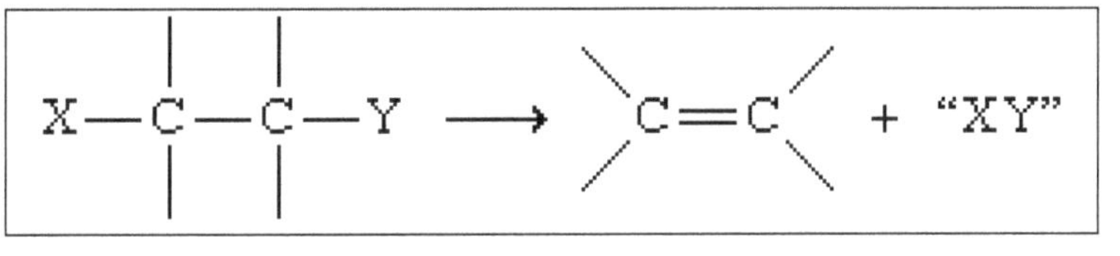

Examples include the dehydration of alcohols:

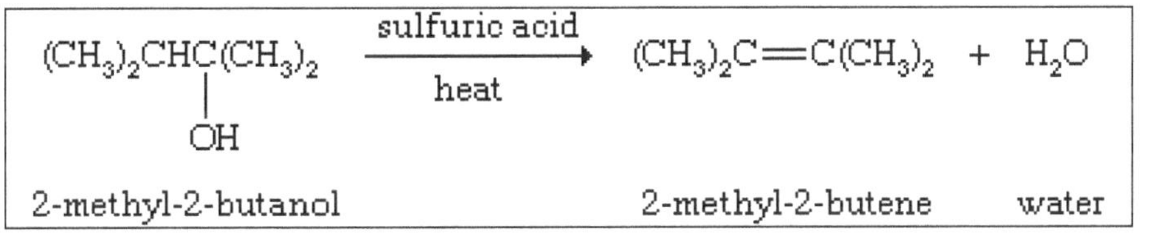

and the dehydrohalogenation (loss of a hydrogen atom and a halogen atom) of alkyl halides.

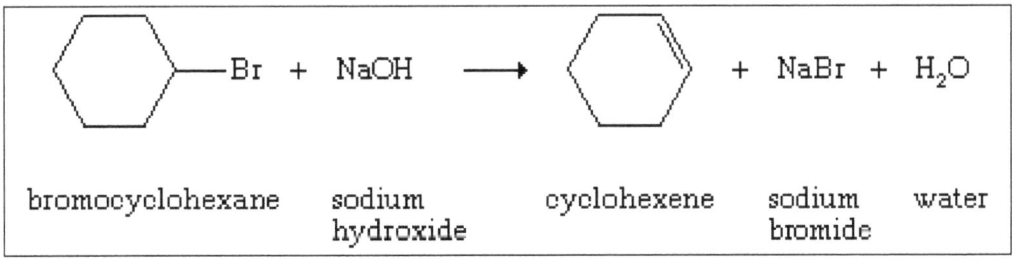

These usually are laboratory rather than commercial methods. Alkenes also can be prepared by partial hydrogenation of alkynes.

Acetylene is prepared industrially by cracking and dehydrogenation of hydrocarbons as described for ethylene. Temperatures of about 800 °C (1,500 °F) produce ethylene; temperatures of roughly 1,150 °C (2,100 °F) yield acetylene. Acetylene, relative to ethylene, is an unimportant industrial chemical. Most of the compounds capable of being derived from acetylene are prepared more economically from ethylene, which is a less expensive starting material. Higher alkynes can be made from acetylene or by double elimination of a dihaloalkane (i.e., removal of both halogen atoms from a disubstituted alkane).

$$CH_3CH_2CH(Br)CH_2Br + 2NaNH_2 \longrightarrow CH_3CH_2C{\equiv}CH + 2NaBr + 2NH_3$$

1,2-dibromobutane — sodium amide — 1-butyne — sodium bromide — ammonia

Physical Properties

The physical properties of alkenes and alkynes are generally similar to those of alkanes or cycloalkanes with equal numbers of carbon atoms. Alkynes have higher boiling points than alkanes or alkenes, because the electric field of an alkyne, with its increased number of weakly held π electrons, is more easily distorted, producing stronger attractive forces between molecules.

Boiling Points of Alkenes And Alkynes		
Name	Formula	Boiling Point (°C)
Ethylene	$CH_2{=}CH_2$	−103.7
Acetylene	HC≡CH	−84.0
Propene	$CH_2{=}CHCH_3$	−47.6
Propyne	$HC{\equiv}CCH_3$	−23.2
1-Butene	$CH_2{=}CHCH_2CH_3$	−6.1
Cis-2-Butene	cis-$CH_3CH{=}CHCH_3$	+3.7
Trans-2-Butene	trans-$CH_3CH{=}CHCH_3$	+0.9
2-Methylpropene	$CH_2{=}C(CH_3)_2$	−6.6
1-Butyne	$HC{\equiv}CCH_2CH_3$	+8.1
2-Butyne	$CH_3C{\equiv}CCH_3$	+27.0
1-Pentene	$CH_2{=}CHCH_2CH_2CH_3$	+30.2
1-Pentyne	$HC{\equiv}CCH_2CH_2CH_3$	+40.2

Chemical Properties

Alkenes react with a much richer variety of compounds than alkanes. The characteristic reaction of alkanes is substitution; that of alkenes and alkynes is addition to the double or triple bond. Hydrogenation is the addition of molecular hydrogen (H_2) to a multiple bond, which converts alkenes to alkanes. The reaction occurs at a convenient rate only in the presence of certain finely divided metal catalysts, such as nickel (Ni), platinum (Pt), palladium (Pd), or rhodium (Rh).

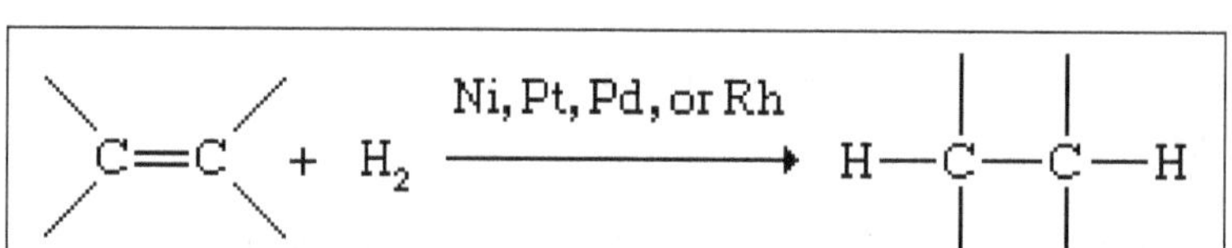

Hydrogenation is used to prepare alkanes and cycloalkanes and also to change the physical properties of highly unsaturated vegetable oils to increase their shelf life. In such processes the liquid

oils are converted to fats of a more solid consistency. Butter substitutes such as margarine are prepared by partial hydrogenation of soybean oil.

Significant progress has been made in developing catalysts for enantioselective hydrogenation. An enantioselective hydrogenation is a hydrogenation in which one enantiomer of a chiral molecule (a molecule that can exist in two structural forms, or enantiomers) is formed in greater amounts than the other. This normally involves converting one of the carbons of the double bond to a stereogenic centre.

Typical catalysts for enantioselective hydrogenation are based on enantiomerically homogeneous ligands bonded to rhodium. Enantioselectivities exceeding 90 percent of a single enantiomer are commonplace in enantioselective hydrogenations, a major application of which is in the synthesis of enantiomerically pure drugs.

The halogens bromine and chlorine add to alkenes to yield dihaloalkanes. Addition is rapid even at room temperature and requires no catalyst. The most important application of this reaction is the addition of chlorine to ethylene to give 1,2-dichloroethane, from which vinyl chloride is prepared.

$CH_2{=}CH_2$	+	Cl_2	$\longrightarrow$	$ClCH_2CH_2Cl$
ethylene		chlorine		1,2-dichloroethane

Compounds of the type HX, where X is a halogen or other electronegative group, also add to alkenes; the hydrogen atom of HX becomes bonded to one of the carbon atoms of the C=C unit, and the X atom becomes bonded to the other.

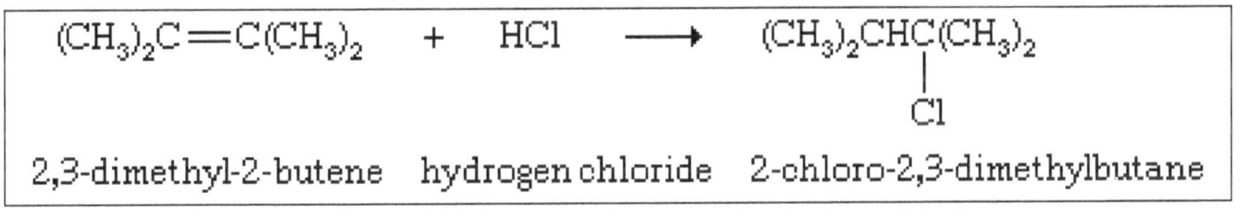

If HX is a strong acid, such as hydrochloric (HCl) or hydrobromic (HBr) acid, the reaction occurs rapidly; otherwise, an acid catalyst is required. One source of industrial ethanol, for example, is the reaction of ethylene with water in the presence of phosphoric acid.

When the two carbon atoms of a double bond are not equivalent, the H of the HX compound adds to the carbon that has the greater number of directly attached hydrogen atoms, and X adds to the one with the fewer. Thus, when sulfuric acid (H_2SO_4) adds to propylene, the product is isopropyl hydrogen sulfate, not n-propyl hydrogen sulfate ($CH_3CH_2CH_2OSO_3H$). This is the first step in the industrial preparation of isopropyl alcohol, which is formed when isopropyl hydrogen sulfate is heated with water.

The term regioselective describes the preference for a reaction that occurs in one direction rather than another, as in the addition of sulfuric acid to propylene. A regiospecific reaction is one that is 100 percent regioselective. The Markovnikov rule expresses the regioselectivity to be expected in the addition of unsymmetrical reagents (such as HX) to unsymmetrical alkenes (such as $H_2C{=}CHR$).

Boron hydrides, compounds of the type R_2BH, add to alkenes to give organoboranes (hydroboration), which can be oxidized to alcohols with hydrogen peroxide (H_2O_2) (oxidation). The net result is the same as if H and –OH add to the double bond with a regioselectivity opposite to the Markovnikov rule. The hydroboration-oxidation sequence is one of a large number of boron-based synthetic methods developed by American chemist Herbert C. Brown.

Vicinal diols, compounds with –OH groups on adjacent carbons, are formed when alkenes react with certain oxidizing agents, especially potassium permanganate ($KMnO_4$) or osmium tetroxide (OsO_4). The most widely used methods employ catalytic amounts of OsO_4 in the presence of oxidizing agents such as tert-butyl hydroperoxide [$(CH_3)_3COOH$].

Alkenes are the customary starting materials from which epoxides, compounds containing a three-membered ring consisting of one oxygen atom and two carbon atoms, are made. The simplest epoxide, ethylene oxide (oxirane), is obtained by passing a mixture of ethylene and air (or oxygen) over a heated silver catalyst. Epoxides are useful intermediates for a number of transformations. Ethylene oxide, for example, is converted to ethylene glycol, which is used in the synthesis of polyester fibres and films and as the main component of automobile antifreeze. On a laboratory scale, epoxides are normally prepared by the reaction of an alkene and a peroxy acid.

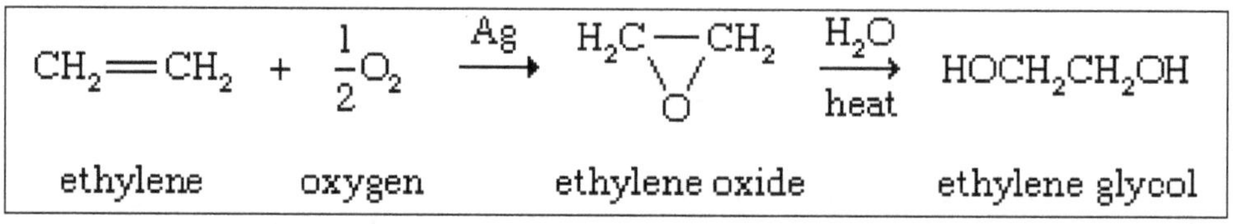

Conjugated dienes undergo a novel and useful reaction known as the Diels-Alder cycloaddition. In this reaction, a conjugated diene reacts with an alkene to form a compound that contains a cyclohexene ring. The unusual feature of the Diels-Alder cycloaddition is that two carbon-carbon bonds are formed in a single operation by a reaction that does not require catalysts of any kind. The German chemists Otto Diels and Kurt Alder received the Nobel Prize for Chemistry in 1950 for discovering and demonstrating the synthetic value of this reaction.

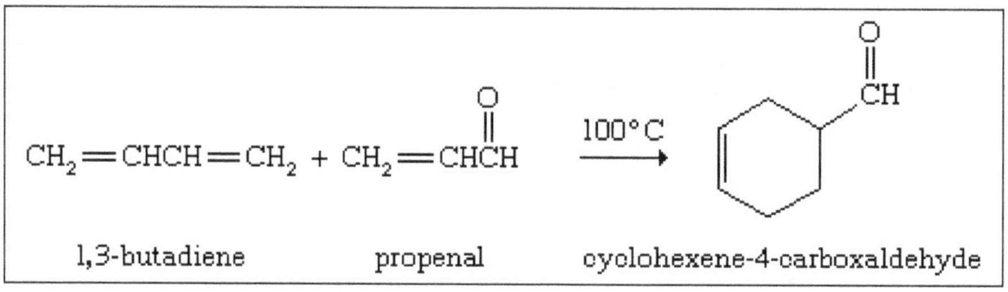

Alkynes undergo addition with many of the same substances that react with alkenes. Hydrogenation of alkynes can be controlled so as to yield either an alkene or an alkane. Two molecules of H_2 add to the triple bond to give an alkane under the usual conditions of catalytic hydrogenation.

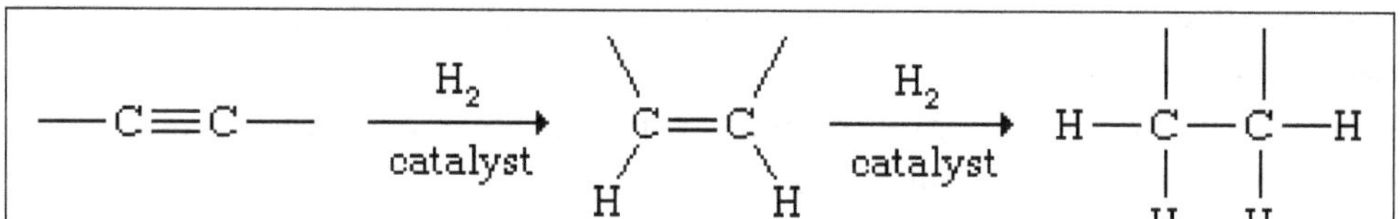

Special, less active (poisoned) catalysts have been developed that permit the reaction to be halted at the alkene stage, and the procedure is used as a method for the synthesis of alkenes. When stereoisomeric alkenes are possible reaction products, the cis isomer is formed almost exclusively.

Alkynes react with Br_2 or Cl_2 by first adding one molecule of the halogen to give a dihaloalkene and then a second to yield a tetrahaloalkane.

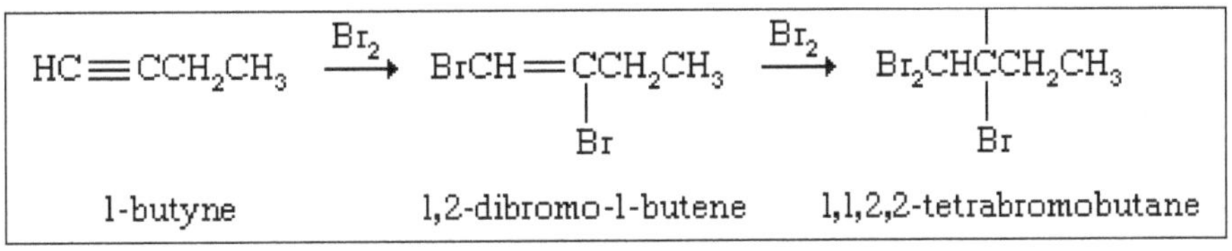

Compounds of the type HX, where X is an electronegative atom or group, also add to alkynes. When acetylene (HC≡CH) reacts with HCl, the product is vinyl chloride (CH_2=CHCl), and, when HCN adds to acetylene, the product is acrylonitrile (CH_2=CHCN). Both vinyl chloride and acrylonitrile are valuable starting materials for the production of useful polymers, but neither is prepared in significant quantities from acetylene, because each is available at lower cost from an alkene (vinyl chloride from ethylene and acrylonitrile from propylene).

Hydration of alkynes is unusual in that the initial product, called an enol and characterized by an H−O−C=C− group, is unstable under the conditions of its formation and is converted to an isomer that contains a carbonyl group.

Although they are very weak acids, acetylene and terminal alkynes are much more acidic than alkenes and alkanes. A hydrogen attached to a triply bonded carbon can be removed by a very strong base such as sodium amide ($NaNH_2$) in liquid ammonia as the solvent.

$$\underset{\text{1-butyne}}{CH_3CH_2C{\equiv}CH} + \underset{\text{sodium amide}}{NaNH_2} \longrightarrow \underset{\text{1-butynylsodium}}{CH_3CH_2C{\equiv}CNa} + \underset{\text{ammonia}}{NH_3}$$

The sodium salt of the alkyne formed in this reaction is not normally isolated but is treated directly with an alkyl halide. The ensuing reaction proceeds with carbon-carbon bond formation and is used to prepare higher alkynes.

$$\underset{\text{1-butynylsodium}}{CH_3CH_2C{\equiv}CNa} + \underset{\text{bromomethane}}{CH_3Br} \longrightarrow \underset{\text{2-pentyne}}{CH_3CH_2C{\equiv}CCH_3} + \underset{\text{sodium bromide}}{NaBr}$$

Polymerization

A single alkene molecule, called a monomer, can add to the double bond of another to give a product, called a dimer, having twice the molecular weight. In the presence of an acid catalyst, the monomer 2-methylpropene (C_4H_8), for example, is converted to a mixture of C_8H_{16} alkenes (dimers) suitable for subsequent conversion to 2,2,4-trimethylpentane (isooctane).

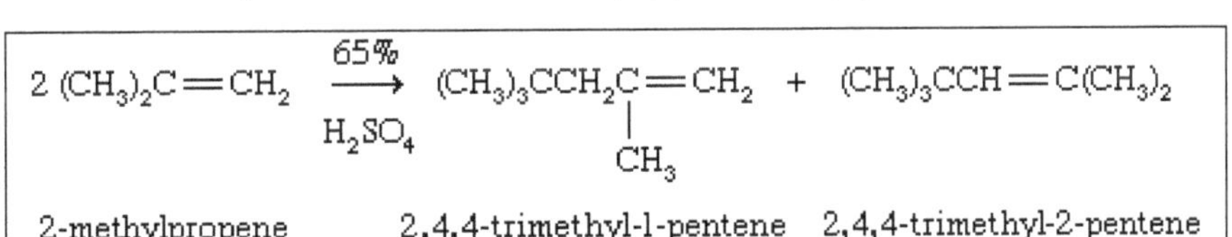

If the process is repeated, trimers, and eventually polymers—substances composed of a great many monomer units—are obtained.

Approximately one-half of the ethylene produced each year is used to prepare the polymer polyethylene. Polyethylene is a mixture of polymer chains of different lengths, where n, the number of monomer units, is on the order of 1,000–5,000.

$$nCH_2{=}CH_2 \xrightarrow{\text{catalyst}} \underset{\text{polyethylene}}{-CH_2CH_2-(CH_2CH_2)_{n-2}-CH_2CH_2-}$$

The distinguishing characteristic of polyethylene is its resistance to attack by most substances. Its resemblance to an alkane in this respect is not surprising, because the polymer chain is nearly void of functional groups. Its ends may have catalyst molecules attached or may terminate in a double bond by loss of a hydrogen atom at the next-to-last carbon. The properties of a particular sample of polyethylene depend mainly on the catalyst used and the conditions under which polymerization occurs. A chain may be continuous, or it may sprout occasional branches of shorter chains. The more nearly continuous the chain, the greater is the density of the polymer.

Low-density polyethylene (LDPE) is obtained under conditions of free-radical polymerization, whereby polymerization is initiated by oxygen or peroxides under high pressure at roughly 200 °C (392 °F). Polyethylene, especially low-density polyethylene, is thermoplastic (softens and flows on heating) and can be extruded into sheets or films and molded into various shapes.

High-density polyethylene (HDPE) is obtained under conditions of coordination polymerization initiated by a mixture of titanium tetrachloride ($TiCl_4$) and triethylaluminum [$(CH_3CH_2)_3Al$]. Coordination polymerization was discovered by German chemist Karl Ziegler. Ziegler and Italian chemist Giulio Natta pioneered the development of Ziegler-Natta catalysts, for which they shared the 1963 Nobel Prize for Chemistry. The original Ziegler-Natta titanium tetrachloride-triethylaluminum catalyst has been joined by a variety of others. In addition to its application in the preparation of high-density polyethylene, coordination polymerization is the method by which ethylene oligomers, called linear α-olefins, and stereoregular polymers, especially polypropylene, are prepared.

Vinyl compounds, which are substituted derivatives of ethylene, can also be polymerized according to the following reaction:

$$\underset{\text{vinyl compound}}{nCH_2{=}CHX} \xrightarrow{\text{catalyst}} \underset{\text{vinyl polymer}}{-CH_2\underset{X}{\underset{|}{C}}H-(CH_2\underset{X}{\underset{|}{C}}H)_{n-2}-CH_2\underset{X}{\underset{|}{C}}H-}$$

Polymerization of vinyl chloride (where X is Cl) gives polyvinyl chloride, or PVC, more than 27 million metric tons of which is used globally each year to produce pipes, floor tiles, siding for houses, gutters, and downspouts. Polymerization of styrene, X = C_6H_5 (a phenyl group derived from benzene), yields polystyrene, a durable polymer used to make luggage, refrigerator casings, and television cabinets and which can be foamed and used as a lightweight packaging and insulating material. If X = CH_3, the product is polypropylene, which is used to make films, molded articles, and fibres. Acrylonitrile, X = CN, gives polyacrylonitrile for use in carpet fibres and clothing.

Diene polymers have an important application as rubber substitutes. Natural rubber is a polymer of 2-methyl-1,3-butadiene (commonly called isoprene). Coordination polymerization conditions have been developed that convert isoprene to a polymer with properties identical to that of natural rubber.

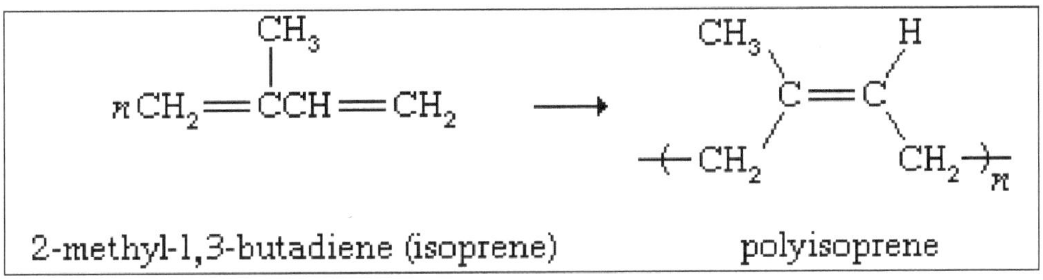

The largest portion of the synthetic rubber industry centres on styrene-butadiene rubber (SBR), which is a copolymer of styrene and 1,3-butadiene. Its major application is in automobile tires.

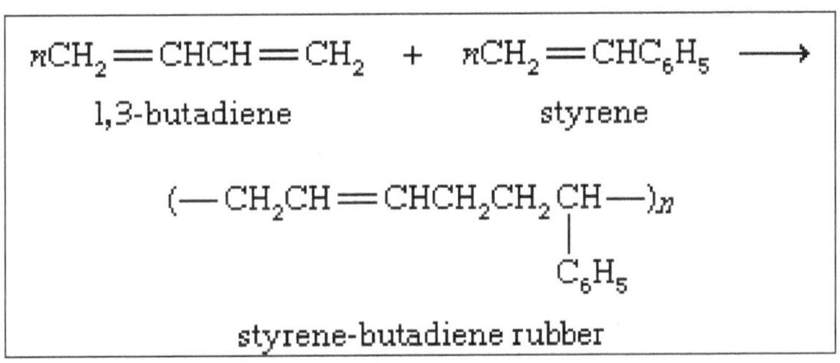

Alkyne polymerization is not nearly as developed nor as useful a procedure as alkene polymerization. The dimer of acetylene, vinylacetylene, is the starting material for the preparation of 2-chloro-1,3-butadiene, which in turn is polymerized to give the elastomer neoprene. Neoprene was the first commercially successful rubber substitute.

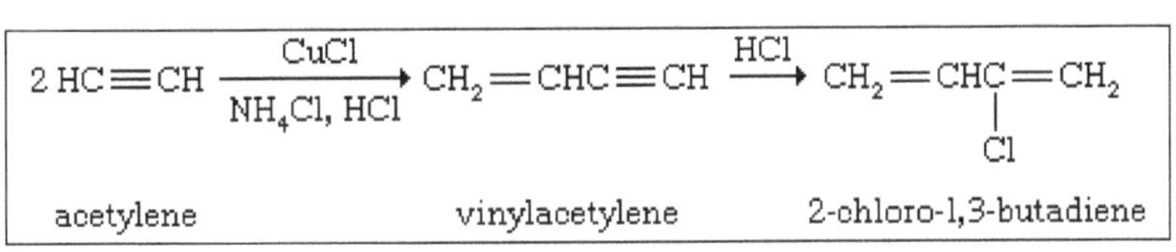

Aromatic Hydrocarbons

Aromatic Hydrocarbons are circularly structured organic compounds that contain sigma bonds along with delocalized pi electrons. They are also referred to as arenes or aryl hydrocarbons.

A few examples of aromatic hydrocarbons are provided below. It can be observed that all these compounds contain a benzene ring.

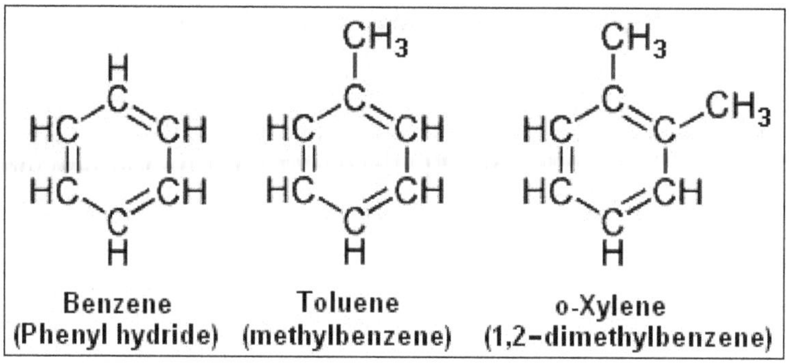

Aromatic Hydrocarbons – Benzene, Toluene and o-Xylene

The aromatic hydrocarbons which do not contain a benzene ring are commonly referred to as heteroarenes. All of these heteroarenes obey Huckel's rule (total number of pi electrons in a monocyclic ring = 4n + 2 where n is any positive integer or zero).

In these types of compounds, a minimum of one carbon is replaced by either nitrogen, oxygen, or sulfur. Common examples of heteroarenes include furan (contains oxygen) and pyridine (contains nitrogen).

Properties of Aromatic Hydrocarbons

The first compound that was categorized as an aromatic hydrocarbon was benzene. It is also the most complex aryl hydrocarbon. Each carbon atom belonging to the benzene ring has two carbon-carbon sigma bonds, one carbon-hydrogen sigma bond, and one double bond with a neighbouring carbon in which the pi electron is delocalized.

This delocalization of pi electrons in the benzene molecule is represented by a circle inside the hexagon. The bond order of all carbon-carbon bonds in this molecule is considered to be 1.5 and this equivalency can be explained with the help of the resonance structures of benzene.

Some general properties of aromatic hydrocarbons have been listed below:

- These compounds exhibit aromaticity (additional stability granted by resonance).
- The ratio of carbon atoms to hydrogen atoms is relatively high in these types of molecules.
- When burnt, the aromatic hydrocarbons display a strong and sooty flame which is yellow in colour.
- These compounds generally undergo electrophilic substitutions and nucleophilic aromatic substitution reactions.

It can be noted that these compounds can be either monocyclic or polycyclic.

Aromatic Hydrocarbons

Benzene (C_6H_6), the simplest aromatic hydrocarbon, was first isolated in 1825 by English chemist Michael Faraday from the oily residues left from illuminating gas. In 1834 it was prepared from benzoic acid ($C_6H_5CO_2H$), a compound obtained by chemical degradation of gum benzoin, the fragrant balsam exuded by a tree that grows on the island of Java, Indonesia. Similarly, the hydrocarbon toluene ($C_6H_5CH_3$) received its name from tolu balsam, a substance isolated from a Central American tree and used in perfumery. Thus benzene, toluene, and related hydrocarbons, while not particularly pleasant-smelling themselves, were classified as aromatic because they were obtained from fragrant substances. Joseph Loschmidt, an Austrian chemist, recognized in 1861 that most aromatic substances have formulas that can be derived from benzene by replacing one or more hydrogens by other atoms or groups. The term aromatic thus came to mean any compound structurally derived from benzene. Use of the term expanded with time to include properties, especially that of special stability, and eventually aromaticity came to be defined in terms of stability alone. The modern definition states that a compound is aromatic if it is significantly more stable than would be predicted on the basis of the most stable Lewis structural formula written for it. All compounds that contain a benzene ring possess special stability and are classified as benzenoid aromatic compounds. Certain other compounds lack a benzene ring yet satisfy the criterion of special stability and are classified as nonbenzenoid aromatic compounds.

Arenes

These compounds are hydrocarbons that contain a benzene ring as a structural unit. In addition to benzene, other examples include toluene and naphthalene.

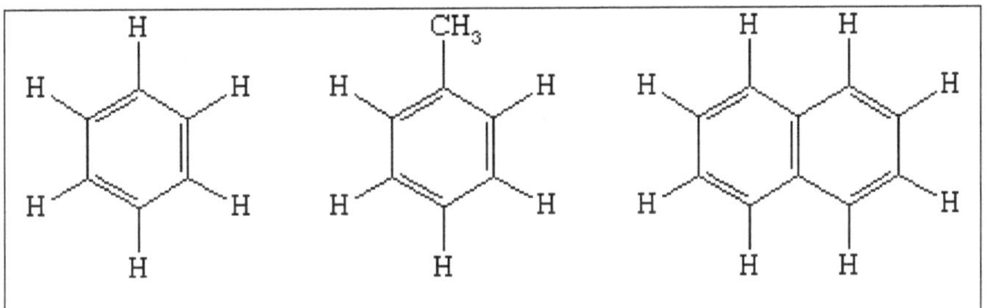

benzene toluene naphthalene

(Hydrogen atoms connected to the benzene ring are shown for completeness in the above structural formulas. The more usual custom, which will be followed hereafter, omits them).

Structure and Bonding

In 1865 the German chemist August Kekule von Stradonitz suggested the cyclic structure for benzene shown above. Kekule's structure, while consistent with the molecular formula and the fact that all of the hydrogen atoms of benzene are equivalent, needed to be modified to accommodate the observation that disubstitution of the ring at adjacent carbons did not produce isomers. Two isomeric products, as shown below, would be expected depending on the placement of the double bonds within the hexagon, but only one 1,2-disubstituted product was formed. In 1872 Kekule revised his proposal by assuming that two such isomers would interconvert so rapidly as to be inseparable from one another.

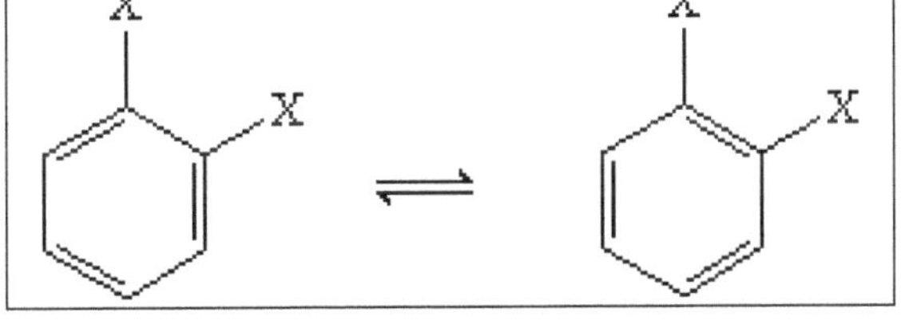

The next major advance in understanding was due largely to the American chemist Linus Pauling, who brought the concept of resonance—which had been introduced in the 1920s—to the question of structure and bonding in benzene. According to the resonance model, benzene does not exist as a pair of rapidly interconverting conjugated trienes but has a single structure that cannot be represented by formulations with localized electrons. The six π electrons (two for the π component of each double bond) are considered to be delocalized over the entire ring, meaning that each π electron is shared by all six carbon atoms rather than by two. Resonance between the two Kekule formulas is symbolized by an arrow of the type ↔ to distinguish it from an interconversion process. The true structure of benzene is described as a hybrid of the two Kekule forms and is often simplified to a hexagon with an inscribed circle to represent the six delocalized π electrons. It is commonly said that a resonance hybrid is more stable than any of the contributing structures, which means, in the case of benzene, that each π electron, because it feels the attractive force of six carbons (delocalized), is more strongly held than if it were associated with only two of them (localized double bonds).

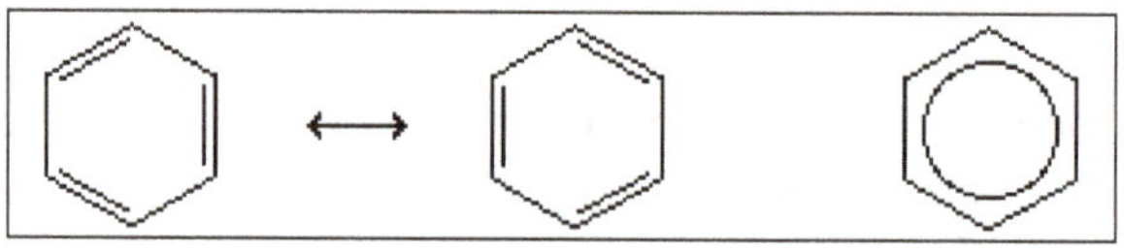

The orbital hybridization model of bonding in benzene is based on a σ bond framework of six sp^2 hybridized carbons. The six π electrons circulate above and below the plane of the ring in a region formed by the overlap of the p orbitals contributed by the six carbons.

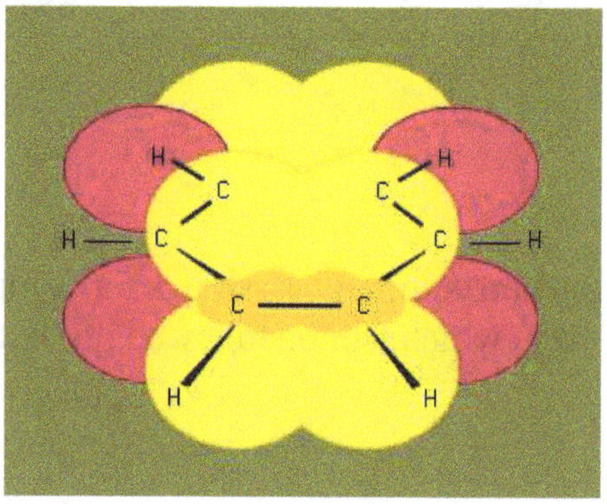

Benzene is the smallest of the organic aromatic hydrocarbons. It contains sigma bonds (represented by lines) and regions of high-pi electron density, formed by the overlapping of p orbitals (represented by the dark yellow shaded area) of adjacent carbon atoms, which give benzene its characteristic planar structure.

Benzene is a planar molecule with six C–C bond distances of equal length. The observed bond distance (1.40 angstroms) is midway between the sp^2-sp^2 single-bond distance (1.46 angstroms) and sp^2-sp^2 double-bond distance (1.34 angstroms) seen in conjugated dienes and is consistent with the bond order of 1.5 predicted by resonance theory. (Bond order is an index of bond strength. A bond order of 1 indicates that a single σ bond exists between two atoms, and a bond order of 2 indicates the presence of one σ and one π bond between two atoms. Fractional bond orders are possible for resonance structures, as in the case of benzene.) Benzene is a regular hexagon; all bond angles are 120°.

The special stability of benzene is evident in several ways. Benzene and its derivatives are much less reactive than expected. Arenes are unsaturated but resemble saturated hydrocarbons (i.e., alkanes) in their low reactivity more than they resemble unsaturated ones. Thermodynamic estimates indicate that benzene is 30–36 kilocalories per mole more stable than expected for a localized conjugated triene structure.

Nomenclature

A number of monosubstituted derivatives of benzene have common names of long standing that have been absorbed into the IUPAC system. Examples include toluene ($C_6H_5CH_3$) and styrene ($C_6H_5CH{=}CH_2$). Disubstituted derivatives of benzene may have their substituents in a 1,2 (ortho, or o), 1,3 (meta, or m), or 1,4 (para, or p) relationship (where the numbers indicate the carbons to which the substituents are bonded) and may be named using either numerical locants or the ortho, meta, para notation.

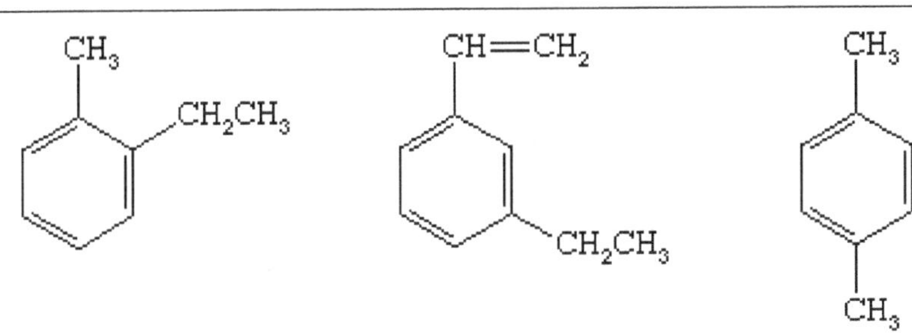

o-ethyltoluene
1-ethyl-2-methylbenzene

m-ethylstyrene
1-ethyl-3-vinylbenzene

p-xylene
1,4-dimethylbenzene

Two groups that contain benzene rings, C_6H_5–(phenyl) and $C_6H_5CH_2$–(benzyl), have special names, as in these examples:

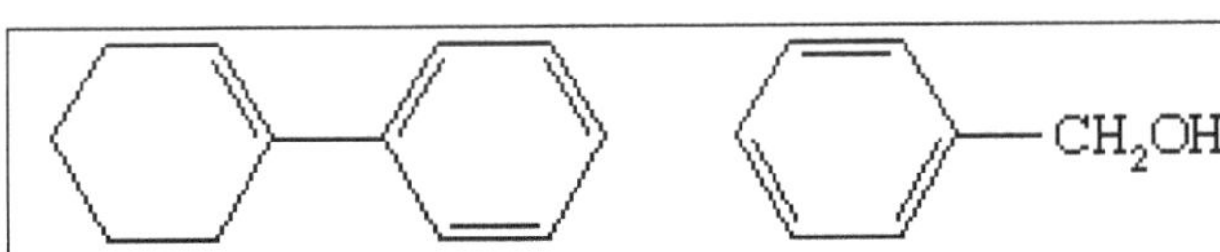

1-phenylcyclohexene

benzyl alcohol

Arenes in which two or more benzene rings share a common side are called polycyclic aromatic compounds. Each such assembly has a unique name, as the examples of naphthalene, anthracene, and phenanthrene illustrate.

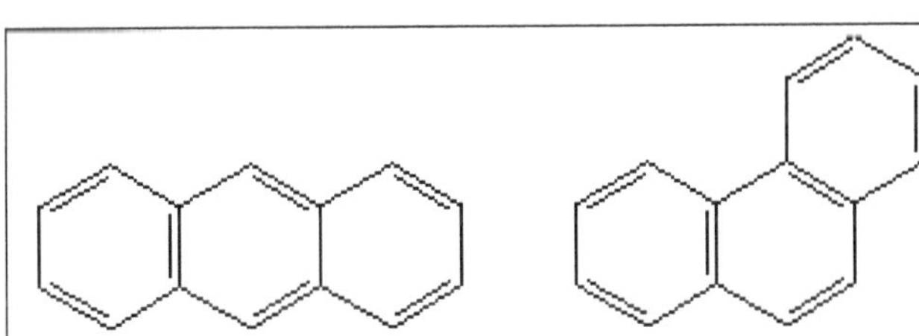

anthracene

phenanthrene

Certain polycyclic aromatic hydrocarbons are known to be carcinogenic and enter the environment when organic matter is burned. Benzo[a]pyrene, for example, is present in tobacco smoke and chimney soot and is formed when meat is cooked on barbecue grills.

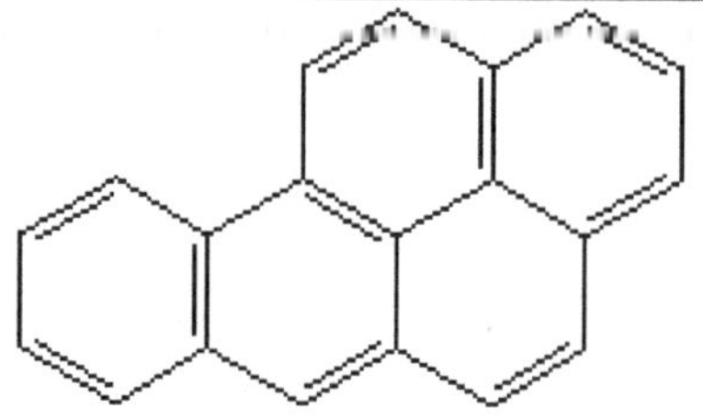

benzo[a]pyrene

Physical Properties

All arenes are either liquids or solids at room temperature; none are gases. Aromatic hydrocarbons are insoluble in water. Benzene was once widely used as a solvent, but evidence of its carcinogenic properties prompted its replacement by less hazardous solvents.

Physical Constants of Benzene and Selected Arenes		
Name	Boiling Point (°C)	Melting Point (°C)
Benzene	80.1	+5.5
Toluene	110.6	−95
Ethylbenzene	136.2	−94
P-Xylene	138.4	+13
Styrene	145	−30.6
Naphthalene	218	+80.3
Anthracene	342	+218
Phenanthrene	340	+100

Source and Synthesis

For a period of approximately 100 years encompassing the last half of the 19th century and the first half of the 20th century, coal was the main starting material for the large-scale production of aromatic compounds. When soft coal is heated in the absence of air, substances are formed that are volatile at the high temperatures employed (500–1,000 °C [930–1,800 °F], depending on the process), which when condensed give the material known as coal tar. Distillation of coal tar gives a number of fractions, the lowest boiling of which contains benzene, toluene, and other low-molecular-weight aromatic compounds. The higher-boiling fractions are sources of aromatic compounds of higher molecular weight. Beginning with the second half of the 20th century, petroleum replaced coal as the principal source of aromatic hydrocarbons. The stability of the benzene ring makes possible processes, known generally as catalytic reforming, in which alkanes are converted to arenes by a combination of isomerization and dehydrogenation events.

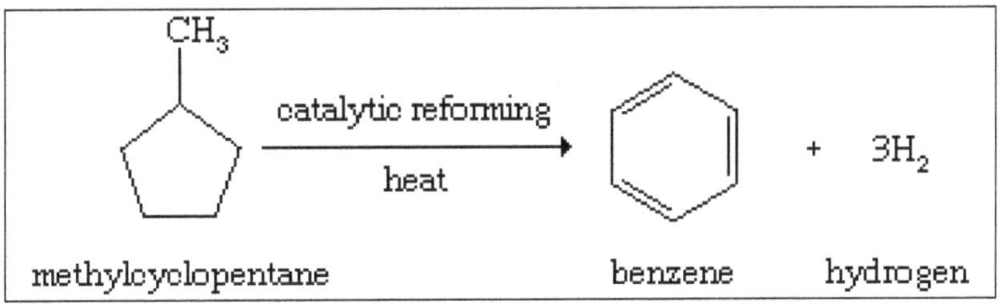

The arenes formed by catalytic reforming are used to boost the octane rating of gasoline and as starting materials for the synthesis of a variety of plastics, fibres, dyes, agricultural chemicals, and drugs.

Reactions

Like other hydrocarbons, arenes undergo combustion to form carbon dioxide and water, and like other unsaturated hydrocarbons, arenes undergo catalytic hydrogenation.

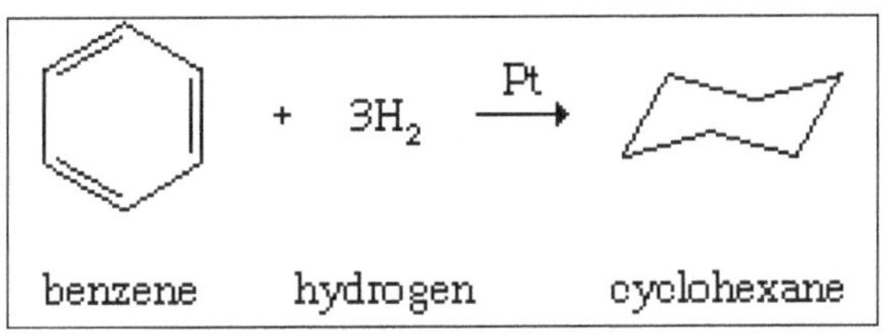

However, many species that react with alkenes by addition react with arenes by replacing one of the hydrogens on the ring (substitution). This behaviour is most pronounced with species known as electrophiles (electron seekers), and the characteristic reaction of an arene is electrophilic aromatic substitution. Representative electrophilic aromatic substitutions, shown with benzene as the arene, include nitration, halogenation, sulfonation, alkylation, and acylation.

nitration:	C_6H_6 benzene	+	$HONO_2$ nitric acid	$\xrightarrow{H_2SO_4}$	$C_6H_5NO_2$ nitrobenzene	+	H_2O water
halogenation: (X = Cl or Br)	C_6H_6 benzene	+	X_2 halogen	$\xrightarrow{FeX_3}$	C_6H_5X halobenzene	+	HX hydrogen halide
sulfonation:	C_6H_6 benzene	+	SO_3 sulfur trioxide	$\longrightarrow$	$C_6H_5SO_3H$ benzenesulfonic acid		
alkylation: (R = alkyl group)	C_6H_6 benzene	+	RCl alkyl halide	$\xrightarrow{AlCl_3}$	C_6H_5R alkylbenzene	+	HCl hydrogen chloride
acylation:	C_6H_6 benzene	+	$R\overset{O}{\overset{\|}{C}}Cl$ acyl chloride	$\xrightarrow{AlCl_3}$	$C_6H_5\overset{O}{\overset{\|}{C}}R$ acylbenzene	+	HCl hydrogen chloride

Alkylation and acylation reactions of aromatic compounds that are catalyzed by aluminum chloride ($AlCl_3$) are referred to as Friedel-Crafts reactions after French chemist and mineralogist Charles

Friedel and American chemist James M. Crafts, who discovered this reaction at the Sorbonne in 1877. Further substitution is possible, and under certain circumstances all six hydrogen atoms of benzene are capable of being replaced. The products of electrophilic aromatic substitution in benzene and its derivatives are employed in subsequent transformations to give a variety of useful products.

The benzene ring is relatively resistant toward oxidation with the exception of its combustion. Arenes that bear alkyl side chains, when treated with strong oxidizing agents, undergo oxidation of the side chain while the ring remains intact.

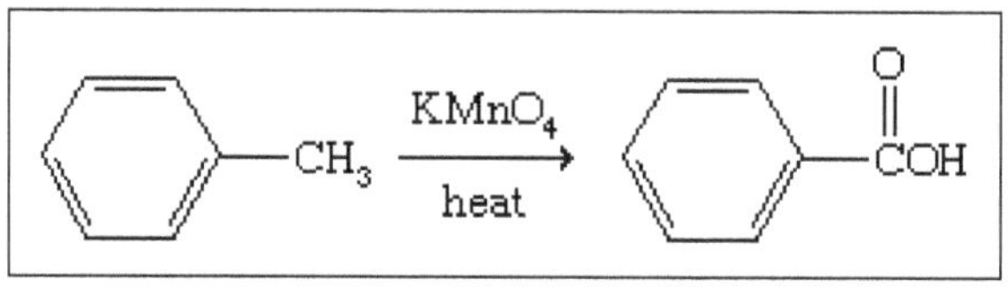

Under conditions of biological oxidation by the cytochrome P-450 enzyme system in the liver, benzene and polycyclic aromatic hydrocarbons undergo epoxidation of their ring. The epoxides that form react with deoxyribonucleic acid (DNA), and it is believed that this process is responsible for the carcinogenic properties of polycyclic aromatic hydrocarbons.

Nonbenzenoid Aromatic Compounds

Once it became clear that the special stability of benzene and related compounds was associated with the cyclic nature of its conjugated system of double bonds, organic chemists attempted to synthesize both larger and smaller analogs. The earliest targets were cyclobutadiene (C_4H_4) and cyclooctatetraene (C_8H_8).

cyclobutadiene (C_4H_4) cyclooctatetraene (C_8H_8)

Of the two, cyclooctatetraene proved more accessible. It was first prepared in 1911 by chemical degradation of the alkaloid pseudopelletierine by the German chemist Richard Willstätter. A more direct synthesis from acetylene was developed in the 1940s. While cyclooctatetraene is a stable substance, thermochemical measurements show that it does not possess the special stability required to be classified as an aromatic hydrocarbon. Structural studies reveal that, unlike benzene in which all of the ring bonds are of equal length (1.40 angstroms), cyclooctatetraene has four short (1.33 angstroms) and four long (1.46 angstroms) carbon-carbon distances consistent with a pattern of alternating single and double bonds. Cyclooctatetraene, moreover, has a nonplanar tub-shaped structure, which, because it is not planar, does not permit the eight π electrons to be delocalized over all the carbon atoms. Classifying cyclooctatetraene as an aliphatic hydrocarbon is also consistent with numerous observations concerning its chemical reactivity, which is characterized by a tendency to undergo alkenelike addition rather than arenelike substitution.

Cyclobutadiene resisted attempts at chemical synthesis until the 1950s when evidence for its intermediacy in certain reactions was obtained. The high reactivity of cyclobutadiene was interpreted as evidence against its aromaticity. Subsequent low-temperature spectroscopic studies revealed that cyclobutadiene has a rectangular structure with alternating single and double bonds unlike the square shape required for an electron-delocalized aromatic molecule.

Annulenes and the Hückel Rule

Insight into the requirements for aromaticity were provided by German physicist Erich Hückel in 1931. Limiting his analysis to planar, monocyclic, completely conjugated polyenes, Hückel calculated that compounds of this type are aromatic if they contain $4n + 2$ π electrons, where n is a whole number. According to the Hückel rule, 2, 6, 10, 14 . . . π electrons ($n = 0, 1, 2, 3$ . . .) confer aromaticity to this class of compounds, but 4, 8, 12, 16 . . . π electrons do not. Benzene, which has 6 π electrons, is aromatic, but cyclobutadiene, which has 4, and cyclooctatraene, which has 8, are nonaromatic.

Monocyclic, completely conjugated polyenes (the hydrocarbons treated by the Hückel rule) are referred to as annulenes, and individual annulenes are differentiated by a numerical prefix equal to the number of π electrons. Beyond benzene,-annulene is the first hydrocarbon to satisfy the Hückel rule. A structure in which all of the double bonds are cis, however, would be a regular 10-sided polygon requiring bond angles of 144° (instead of the 120° angles required for sp^2 hybridized carbon) and would suffer considerable angle strain. The destabilization owing to angle strain apparently exceeds the stabilization associated with aromaticity and makes all-cis-cyclodecapentaene a highly reactive substance. An isomer in which two of the double bonds are trans should, in principle, be free of angle strain. It is destabilized, however, by a repulsive force between two hydrogen atoms that are forced together in the interior of the ring, and for this reason it is relatively reactive.

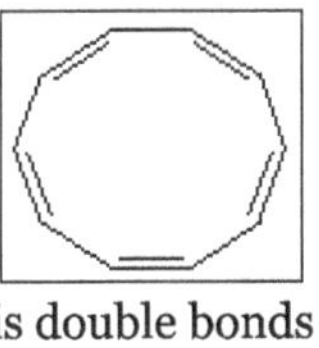

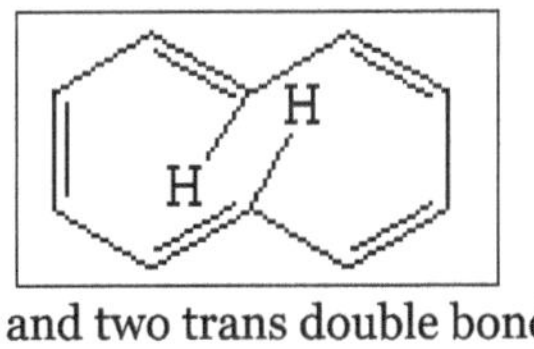

Five cis double bonds three cis and two trans double bonds
Cyclodecapentaene stereoisomers

[18]-Annulene is predicted to be aromatic by the Hückel rule ($4n + 2 = 18$ when $n = 4$). The structure shown has a shape that makes it free of angle strain and is large enough so that repulsive forces between hydrogen atoms in the interior are minimal. Thermochemical measurements indicate a resonance energy of roughly 100 kilocalories per mole, and structural studies reveal that the molecule is planar with all its bond distances falling in the range 1.37–1.43 angstroms. In terms of its chemical reactivity, however, [18]-annulene resembles an alkene more than it resembles benzene.

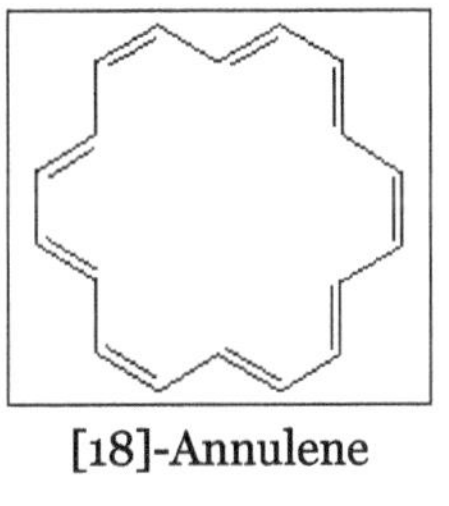

[18]-Annulene

Polycyclic Nonaromatic Compounds

The Hückel rule is not designed to apply to polycyclic compounds. Nevertheless, a similar dependence on the number of π electrons is apparent. The bicyclic hydrocarbon azulene has the same number of π electrons (10) as naphthalene and, like naphthalene, is aromatic. Pentalene and heptalene, analogs with 8 and 12 π electrons, respectively, are not aromatic. Both are relatively unstable, highly reactive substances.

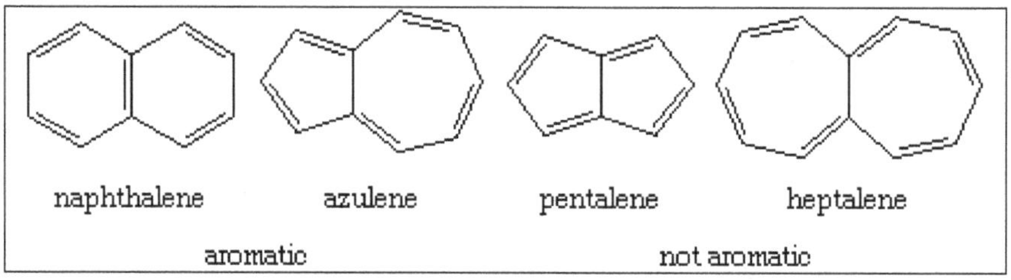

Organometallic Compounds

Organometallic compound is any member of a class of substances containing at least one metal-to-carbon bond in which the carbon is part of an organic group. Organometallic compounds constitute a very large group of substances that have played a major role in the development of the science of chemistry. They are used to a large extent as catalysts (substances that increase the rate of reactions without themselves being consumed) and as intermediates in the laboratory and in industry. The class includes such compounds as ferrocene, a remarkably stable compound in which an iron atom is sandwiched between two hydrocarbon rings.

Organometallic compounds are typically discussed in terms of the metal as either main-group compounds or transition metal compounds. The main-group metals of organometallic compounds are typically considered to be those of the S-block (groups 1 and 2) and the heavier elements of the p-block (groups 13–15) in the periodic table of elements. The transition metals include those elements in the d- and f-blocks (groups 3–12).

The physical and chemical properties of organometallic compounds vary greatly. Most are solids, particularly those whose hydrocarbon groups are ring-shaped or aromatic, but some are liquids and some are gases. Their heat and oxidation stability vary widely. Some are very stable, but a number of compounds of electropositive elements such as lithium, sodium, and aluminum are spontaneously flammable. Many organometallic compounds are highly toxic, especially those that are volatile.

The properties of the organometallic compounds depend in large measure on the type of carbon-metal bonds involved. Some are ordinary covalent bonds, in which pairs of electrons are shared between atoms. Others are multicentre covalent bonds, in which the bonding involves more than two atoms. A third type are ionic bonds, in which the bonding electron pair is donated by only one atom. In donor-acceptor bonds, the metal atom is connected to hydrocarbons with multiple bonds between carbon atoms.

Where metal atoms form covalent bonds with carbon atoms, the electrons are usually shared unequally. As a result, the bond is polarized—one end is more negative than the other. The extent of polarization depends on the strength with which the metal atom binds electrons. Organometallic compounds range in polar power from methylpotassium, in which the bond is almost like certain ionic bonds, to lead, which bonds with carbon with very little polarization.

Importance of Organometallic Compounds

Because of bond polarity, many organometallic compounds have reactivities that have made them important in chemical synthesis. The organomagnesium halides (Grignard reagents), for example, are used widely in synthetic organic chemistry, as are organolithium and organoboron compounds. Alkylaluminum compounds are also employed in organic synthesis. Used with titanium salts, they are important catalysts in the polymerization of unsaturated hydrocarbons, such as ethylene and propylene. The mechanism of action of the titanium-aluminum alkyl catalysts probably involves interaction between the titanium atoms and the double bonds of the hydrocarbons.

Organometallic compounds containing lead, tin, and mercury are all commercially significant. A large number of organotin compounds, for example, are used as pharmaceuticals, pesticides, stabilizers for polyvinyl chloride, and fire retardants. Methylmercury has caused severe pollution problems as a result of its toxicity. This fact has led to stringent controls on the discharge of mercury from chemical plants into rivers, lakes, and oceans.

Carbon monoxide reacts readily with many transition-metal atoms to form metal carbonyls, themselves a class of organometallics. One of the earliest to be discovered was tetracarbonylnickel, a volatile nickel compound that became the basis of a process for purifying nickel. Metal carbonyls are employed as catalysts in many reactions in the petrochemical industry.

Properties of Organometallic Compounds

A few properties of organometallic compounds are listed below as short points:

- The bond between the metal and the carbon atom is often highly covalent in nature.
- Most of the organometallic compounds exist in solid states, especially the compounds in which the hydrocarbon groups are aromatic or have a ring structure.
- The compounds consisting of highly electropositive metals such as sodium or lithium are very volatile and can undergo spontaneous combustion.
- In many cases, organometallic compounds are found to be toxic to humans (especially the compounds that are volatile in nature).
- These compounds can act as reducing agents, especially the compounds formed by highly electropositive metals.

From the points given above, it can be observed that the properties of organometallic compounds differ amongst each other based on the properties of the metals that constitute them.

Main Group Organometallic Chemistry

Alkali and Akaline Earths Organometallic

- Li, Na, K organyls

- Be organyls
- Mg organyls

Aluminium Group

Silicon Group

- Si organyls
- Ga organyls

Pb, Sb, Sn and Hg

Transition-metal Organometallic Chemistry

The organometallic chemistry of the transition elements is quite different from the main-group ones due to the availability for bonding of the *n d* orbitals with consequent ability for the central atom to change geometry and expand the octet.

Single σ-bonding

- M-alkyl
- β-elimination

π-acceptor Bonding

- Alkene Complexes

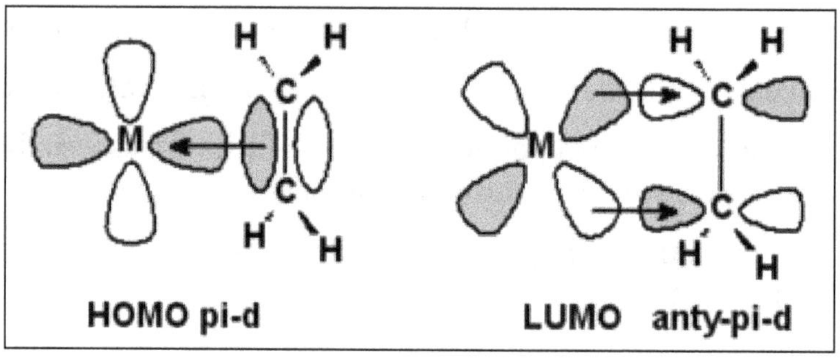

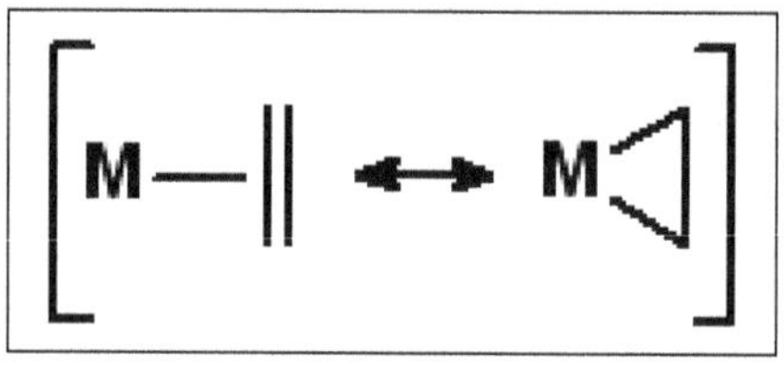

- CO Complexes
    - $\sigma^* \rightarrow d_\sigma$
    - $d_\pi \rightarrow \pi^*$

These interactions are synergicin increading the M-CO bond strength. In fact, the second interaction, as known as *pi* backbonding increases the available electron density on the CO.

The partial filling of the π^* orbital leads to a weakened C-O triple bond, as showed from the stretching frequencies (in cm^{-1}) of CO free and in M/CO complexes.

free CO	2143
$V(CO)_6$	1976
$Ni(CO)_4$	2057
$Cr(CO)_6$	2000

- Arene Complexes
- Fischer Carbenes

  By treatment of a CO complex with a strong nucleophile

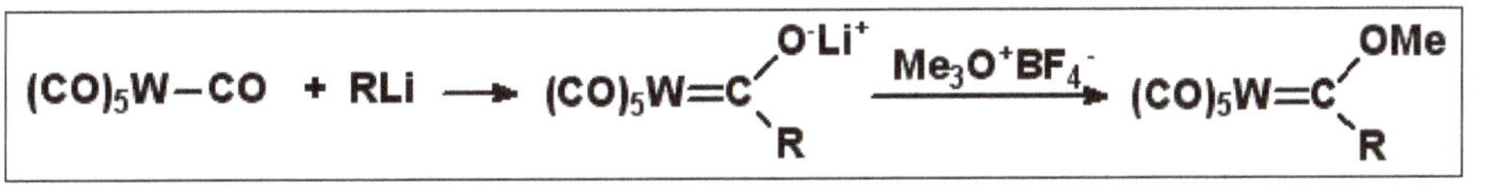

- Schrock Carbenes

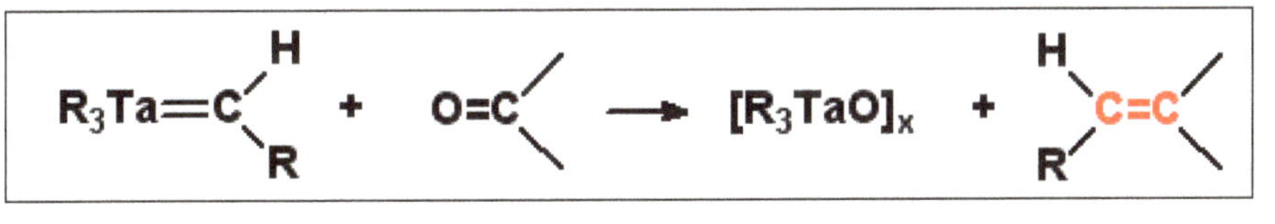

Applications

Organometallic Compounds have a broad range of applications in the field of chemistry. Some of them are given below:

- In some of the commercial chemical reactions, organometallic compounds are used as homogeneous catalysts.
- These compounds are used as stoichiometric reagents in both industrial and research-oriented chemical reactions.
- These compounds are also used in the manufacture of some of the semiconductors, which require the use of compounds such as trimethylgallium, trimethylaluminum, trimethylindium, and trimethyl antimony.
- They are also used in the production of light emitting diodes (or LEDs).
- These compounds are employed in bulk hydrogenation processes such as the production of margarine.
- These compounds are used as catalysts and reagents during the synthesis of some organic compounds.

- The complexes formed from organometallic compounds are useful in the facilitation of the synthesis of many organic compounds.

Organometallic Compounds in Catalysis

Catalysts are substances that increase the rate of a reaction but are not consumed in the reaction. Catalysts are widely encountered in nature, industry, and the laboratory. Many of the catalysts utilized in the chemical industry and the laboratory are organometallic compounds.

Hydrogenation

The overall result of the catalytic hydrogenation of alkenes is to add molecular hydrogen, H_2, across the double bond of an alkene. The reactants, H_2 and ethylene (C_2H_4), enter the cycle by reaction with the complex to produce in succession a hydrido complex and an alkene complex. In the final step, the hydrogenated product leaves the loop with the regeneration of the coordinatively unsaturated Rh complex. The cycle continues as long as hydrogen gas and ethylene are supplied. The rhodium complexes in solution are the catalysts and are not used up in the reaction. Rhodium, which is more expensive than platinum, can be used even in catalytic processes where the products are inexpensive, because the rhodium is not consumed. Modifications of this type of catalyst are also employed in the production of pharmaceuticals such as levodopa (or L-dopa), which is used to treat Parkinson disease.

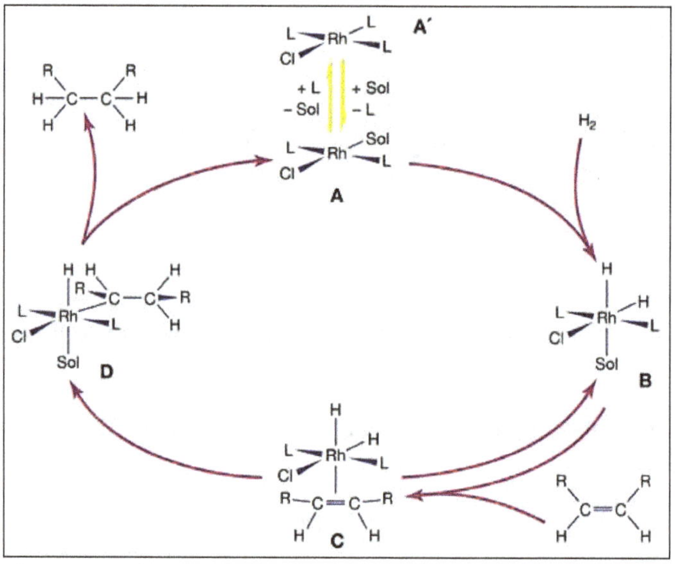

There are multiple steps in the catalytic hydrogenation of alkenes (RHC=CRH compounds) by a rhodium (Rh) complex. L is the triphenylphosphine ligand, PPh_3, and Sol is a solvent molecule.

Hydroformylation

Hydroformylation involves the addition of carbon monoxide and hydrogen to an alkene to form an aldehyde containing one more carbon atom than the original alkene.

$$R{-}CH{=}CH_2 + CO + H_2 \longrightarrow R{-}CH_2{-}CH_2{-}\overset{\|}{C}{-}H$$

This catalytic reaction is employed in the petrochemical industry, where $Co_2(CO)_8$ or various rhodium catalysts are utilized. The catalytic cycle proceeds through a series of organometallic intermediates. The aldehydes produced by hydroformylation are normally reduced to alcohols that are used as solvents, as plasticizers, and in the synthesis of detergents. The scale of production is enormous, amounting to millions of tons per year.

Alkene Polymerization

Polyalkenes, the most common and useful class of synthetic polymers, are often prepared by use of organometallic catalysts, either in solution or supported on a solid surface. In the 1950s, the German chemist Karl Ziegler developed a catalyst for ethylene polymerization based on a catalyst formed by the reaction of $TiCl_4$ with $Al(C_2H_5)_3$. Soon thereafter, Italian chemist Giulio Natta made use of this type of catalyst for the polymerization of propylene to produce polymers with highly regular structures. The intimate details of the reactions of these commercial catalytic processes are not entirely understood, but there are strong indications from more easily studied soluble organometallic catalysts that alkenes coordinate to a metal centre and then insert into a hydrocarbon chain, producing a longer-chain hydrocarbon attached to the metal centre. Repetition of this process leads to extremely-long-chain hydrocarbon polymers, which include many of the most familiar plastics, such as polyethylene and polypropylene. These plastics are used in consumer items ranging from milk containers and plastic bags to artificial limbs and car bumpers.

Heterocyclic Compounds

Heterocyclic compound is also called heterocycle, any of a major class of organic chemical compounds characterized by the fact that some or all of the atoms in their molecules are joined in rings containing at least one atom of an element other than carbon (C). The cyclic part of heterocyclic indicates that at least one ring structure is present in such a compound, while the prefix hetero- refers to the noncarbon atoms, or heteroatoms, in the ring. In their general structure, heterocyclic compounds resemble cyclic organic compounds that incorporate only carbon atoms in the rings—for example, cyclopropane (with a three-carbon-atom ring) or benzene (with a six-carbon-atom ring)—but the presence of the heteroatoms gives heterocyclic compounds physical and chemical properties that are often quite distinct from those of their all-carbon-ring analogs.

Heterocyclic compounds include many of the biochemical material essential to life. For example, nucleic acids, the chemical substances that carry the genetic information controlling inheritance, consist of long chains of heterocyclic units held together by other types of materials. Many naturally occurring pigments, vitamins, and antibiotics are heterocyclic compounds, as are most hallucinogens. Modern society is dependent on synthetic heterocycles for use as drugs, pesticides, dyes, and plastics.

General aspects of Heterocyclic Compounds

The most common heterocycles are those having five- or six-membered rings and containing heteroatoms of nitrogen (N), oxygen (O), or sulfur (S). The best known of the simple heterocyclic compounds are pyridine, pyrrole, furan, and thiophene. A molecule of pyridine contains a ring of six atoms—five carbon atoms and one nitrogen atom. Pyrrole, furan, and thiophene molecules each contain five-membered rings, composed of four atoms of carbon and one atom of nitrogen, oxygen, or sulfur, respectively.

Pyridine and pyrrole are both nitrogen heterocycles—their molecules contain nitrogen atoms along with carbon atoms in the rings. The molecules of many biological materials consist in part of pyridine and pyrrole rings, and such materials yield small amounts of pyridine and pyrrole upon strong heating. In fact, both of these substances were discovered in the 1850s in an oily mixture formed by strong heating of bones. Today, pyridine and pyrrole are prepared by synthetic reactions. Their chief commercial interest lies in their conversion to other substances, chiefly dyestuffs and drugs. Pyridine is used also as a solvent, a waterproofing agent, a rubber additive, an alcohol denaturant, and a dyeing adjunct.

Furan is an oxygen-containing heterocycle employed primarily for conversion to other substances (including pyrrole). Furfural, a close chemical relative of furan, is obtained from oat hulls and corncobs and is used in the production of intermediates for nylon. Thiophene, a sulfur heterocycle, resembles benzene in its chemical and physical properties. It is a frequent contaminant of the benzene obtained from natural sources and was first discovered during the purification of benzene. Like the other compounds, it is used primarily for conversion to other substances. Furan and thiophene were both discovered in the latter part of the 19th century.

In general, the physical and chemical properties of heterocyclic compounds are best understood by comparing them with ordinary organic compounds that do not contain heteroatoms.

Comparison with Carbocyclic Compounds

The molecules of organic chemical compounds are built up from a framework or backbone of carbon atoms to which are attached hydrogen (H), oxygen, or other heteroatoms. Carbon atoms have the unique property of being able to link with one another to form chains of atoms. When the ends of the chains are joined together into a ring, cyclic compounds result; such substances often are referred to as carbocyclic or alicyclic compounds. Substitution of one or more of the ring carbon atoms in the molecules of a carbocyclic compound with a heteroatom gives a heterocyclic compound.

A typical carbocyclic compound is cyclopentane (C_5H_{10}), the molecular structure of which is indicated by the formula given below in which the chemical symbols represent atoms of the elements and the lines represent bonds between the atoms.

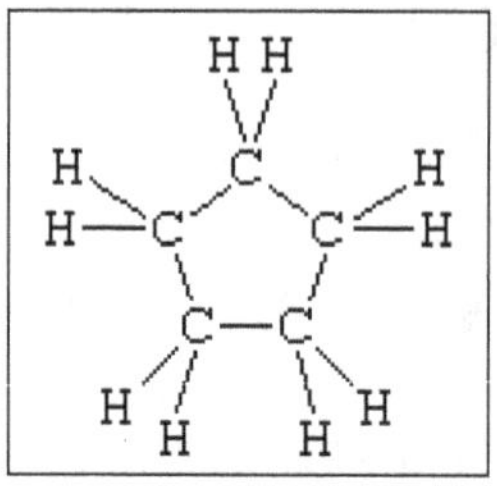

For convenience such formulas are often written in simplified polygonal form, such as the structure given below.

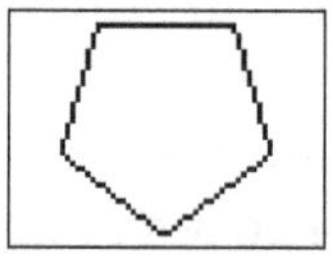

For cyclopentane, in which each corner of the polygon represents a carbon atom (it being understood that hydrogen atoms are joined to the carbon atoms as required).

When one of the carbon atoms of cyclopentane is replaced with an atom of nitrogen, the compound pyrrolidine, a chemical relative of pyrrole, is produced. The structural formula of pyrrolidine is written below.

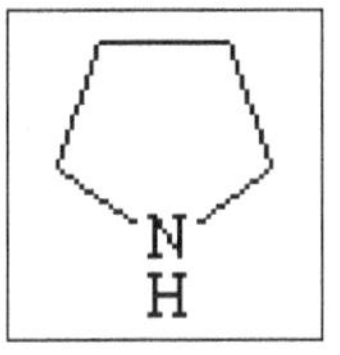

Other heterocyclic compounds can be envisioned similarly as derivatives of cyclopentane by substitution with other heteroatoms or of other carbocyclic compounds by substitution with nitrogen or other heteroatoms.

The simplest organic compounds are the hydrocarbons, compounds of carbon and hydrogen only. Hydrocarbons are classified as saturated if all four possible bonds of every carbon atom are joined singly to another carbon atom or to a hydrogen atom. They are classified as unsaturated if they contain a double or triple bond between any two of the carbon atoms, and they are classed as aromatic if they contain at least one ring, all atoms of which are joined by alternating double and single bonds. Nonaromatic unsaturated compounds are highly reactive—that is, they readily undergo additions of atoms or groups of atoms to the carbon atoms of their double bonds, giving each carbon four substituents. Aromatic compounds, though having double bonds, are extremely stable and do not readily undergo the addition reactions characteristic of other unsaturated compounds. The stability and unreactivity of, for instance, a six-membered aromatic ring are associated with the presence of three pairs of electrons, called pi (π) electrons, associated with the three double bonds of the ring. Together these electrons, constituting the so-called aromatic sextet, form an unusually stable structure associated with the aromatic ring as a whole rather than with the individual pairs of atoms.

Heterocycles too may be classified as saturated, unsaturated, or aromatic. Thus, as the following structural formulas, pyrrolidine is a saturated heterocyclic compound containing no double bond; 4,5-dihydrofuran is an unsaturated heterocyclic compound; and pyridine is a typical heterocyclic aromatic, or heteroaromatic, substance. In the two structural formulas given for pyridine, the first shows the double bonds, whereas the second represents the aromatic sextet with a circle.

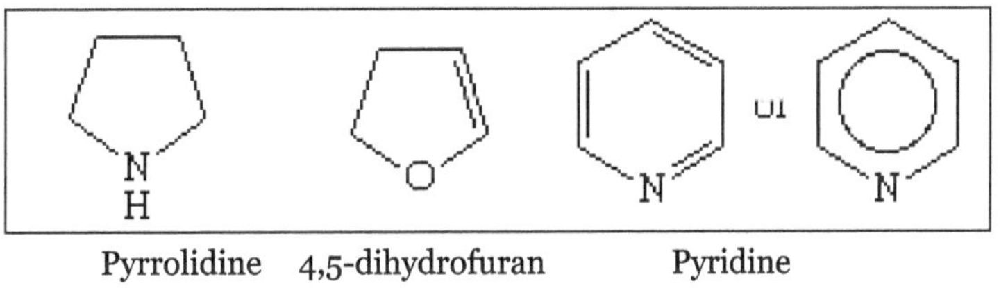

Pyrrolidine 4,5-dihydrofuran Pyridine

This classification relates the chemistry of heterocycles directly with that of carbocyclic derivatives, which are usually better known. In general, synthetic methods and physical and chemical properties of the saturated and the partly unsaturated heterocyclic compounds closely resemble those for their acyclic (noncyclic, or open-chain) analogs. Thus, pyrrolidine may be considered as a cyclic secondary amine and has much in common with the corresponding acyclic amine, diethylamine, which is represented by the formula:

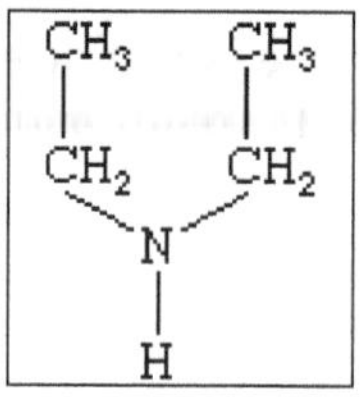

Similarly, 4,5-dihydrofuran mirrors many of the properties of the corresponding unsaturated ether, ethyl vinyl ether, which has the formula:

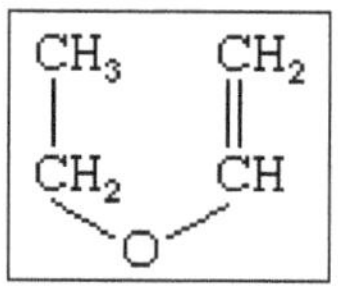

Nomenclature of Heterocyclic Compounds

Naming heterocyclic compounds is complicated because of the existence of many common names in addition to the internationally agreed-upon systematic nomenclature.

The types of heteroatoms present in a ring are indicated by prefixes; in particular, oxa-, thia-, and aza- denote oxygen, sulfur, and nitrogen atoms, respectively. The numbers of heteroatoms of a particular kind are indicated by number prefixes joined to the heteroatom prefixes, as dioxa- and triaza-. The presence of different kinds of heteroatoms is indicated by combining the above prefixes, using the following order of preference: oxa- first, followed by thia- and then aza-.

In addition, partially saturated rings are indicated by the prefixes dihydro-, tetrahydro-, and so on, according to the number of "extra" hydrogen atoms bonded to the ring atoms. The positions of heteroatoms, extra hydrogen atoms, and substituents are indicated by Arabic numerals, for which the numbering starts at an oxygen atom, if one is present, or at a sulfur or nitrogen atom and continues in such a way that the heteroatoms are assigned the lowest possible numbers. Other things being equal, numbering starts at a nitrogen atom that carries a substituent rather than at multiply bonded nitrogen. In compounds with maximum unsaturation, if the double bonds can be arranged in more than one way, their positions are defined by indicating the nitrogen or carbon atoms that are not multiply bonded and that consequently carry an extra hydrogen atom (or substituent), as follows: 1H-, 2H-, and so on.

Nature of Heteroaromaticity

Aromaticity denotes the significant stabilization of a ring compound by a system of alternating single and double bonds—called a cyclic conjugated system—in which six π electrons generally participate. A nitrogen atom in a ring can carry a positive or a negative charge, or it can be in the neutral form. An oxygen or sulfur atom in a ring can either be in the neutral form or carry a positive charge. A fundamental distinction is usually made between (1) those heteroatoms that participate in a cyclic conjugated system by means of a lone, or unshared, pair of electrons that are in an orbital perpendicular to the plane of the ring and (2) those heteroatoms that do so because they are connected to another atom by means of a double bond.

An example of an atom of the first type is the nitrogen atom in pyrrole, which is linked by single covalent bonds to two carbon atoms and one hydrogen atom. Nitrogen has an outermost shell of five electrons, three of which can enter into three covalent bonds with other atoms. After the bonds are formed, as in the case of pyrrole, there remains an unshared electron pair that can engage in cyclic conjugation. The aromatic sextet in pyrrole is made up of two electrons from each of the two carbon-carbon double bonds and the two electrons that compose the unshared electron pair of the nitrogen atom. As a consequence, there tends to be a net flow of electron density from the nitrogen atom to the carbon atoms as the nitrogen's electrons are drawn into the aromatic sextet. Alternatively, the pyrrole molecule may be described as a resonance hybrid—that is, a molecule whose true structure can only be approximated by two or more different forms, called resonance forms.

An example of a heteroatom of the second type is the nitrogen atom in pyridine, which is linked by covalent bonds to only two carbon atoms. Pyridine also has a π-electron sextet, but the nitrogen atom contributes only one electron to it, one additional electron being contributed by each of the five carbon atoms in the ring. In particular, the unshared electron pair of the nitrogen atom is not involved. Moreover, because nitrogen's attraction for electrons (its electronegativity) is greater than that of carbon, electrons tend to move toward the nitrogen atom rather than away from it, as in pyrrole.

Quite generally, heteroatoms may be referred to as pyrrolelike or pyridine-like, depending on whether they fall into the first or second class. The pyrrolelike heteroatoms $-NR-$ (R being hydrogen or a hydrocarbon group), $-N^{-}-$, $-O-$, and $-S-$ tend to donate electrons into the π-electron system, whereas the pyridine-like heteroatoms $-N=$, $-N^{+}R=$, $-O^{+}=$, and $-S^{+}=$ tend to attract the π electrons of a double bond.

In six-membered heteroaromatic rings, the heteroatoms (usually nitrogen) are pyridine-like—for example, the compounds pyrimidine, which contains two nitrogen atoms, and 1,2,4-triazine, which contains three nitrogen atoms.

Six-membered heteroaromatic compounds cannot normally contain pyrrole like heteroatoms. Five-membered heteroaromatic rings, however, always contain one pyrrolelike nitrogen, oxygen, or sulfur atom, and they may also contain up to four pyridine-like heteroatoms, as in the compounds thiophene (with one sulfur atom), 1,2,4-oxadiazole (with one oxygen atom and two nitrogen atoms), and pentazole (with five nitrogen atoms).

The quantitative measurement of aromaticity—and even its precise definition—has challenged chemists since German chemist August Kekule formulated the ring structure for benzene in the mid-19th century. Various methods based on energetic, structural, and magnetic criteria have been widely used to measure the aromaticity of carbocyclic compounds. All of them, however, are difficult to apply quantitatively to heteroaromatic systems because of complications arising from the presence of heteroatoms.

Chemical reactivity can provide a certain qualitative insight into aromaticity. The reactivity of an aromatic compound is affected by the extra stability of the conjugated system that it contains; the extra stability in turn determines the tendency of the compound to react by substitution of hydrogen—i.e., replacement of a singly bonded hydrogen atom with another singly bonded atom or group—rather than by addition of one or more atoms to the molecule via the breaking of a double bond. In terms of reactivity, therefore, the degree of aromaticity is measured by the relative tendency toward substitution rather than addition. By this criterion, pyridine is more aromatic than furan, but it is difficult to say just how much more aromatic.

Physical Properties of Heterocyclic Compounds

Physical properties are important as criteria for judging the purity of heterocycles just as for other organic compounds. Organic compounds generally show great regularity in their physical properties, and heterocycles are no exception.

The melting point was once a widely used criterion for purity, but it has been increasingly superseded by optical spectra, based on light absorption; mass spectra, based on relative masses of molecular fragments; and magnetic resonance spectra, based on nuclear properties. Nevertheless, knowledge of melting and boiling points is still helpful for judging the purity of a compound.

Melting and Boiling Points

The boiling points of certain saturated heterocycles are listed in the first table and are compared with those of the corresponding cycloalkanes (rightmost column of the table).

Boiling Points (°C) of Saturated Heterocycles and Corresponding Carbocycles					
Ring Size	Number (and Position) of Heteroatoms	Type of Heteroatom			Saturated Cycloalkane
		N (as NH)	O	S	
3	one	56	11	55	−33
4	one	63	48	94	13
5	one	87	65	121	49
6	one	106	88	141	80
6	two (1,2)	150	116	190	80
6	two (1,3)	150	106	207	80
6	two (1,4)	145	101	200	80
7	one	138	120	174	119

Replacement of a two-carbon unit (two carbon and two hydrogen atoms, molecular weight equal to 26) by a single sulfur atom (atomic weight 32) has little effect on the melting or boiling point. On the other hand, replacement of a two-carbon unit by an oxygen atom (atomic weight 16) lowers the boiling point by about 40 °C (72 °F), which is to be expected because of the decreased molecular weight of the furan compounds (lighter compounds being more volatile). Introduction of nitrogen atoms into the benzene ring is accompanied by less-regular changes. Replacement of a two-carbon unit by an imino (NH) group, or of a single carbon by a nitrogen atom, increases the boiling point. Furthermore, making these two changes simultaneously increases the boiling point even more, probably as the result of intermolecular association by hydrogen bonding (a weak form of attachment via certain types of hydrogen atoms) between the pyridine-like nitrogen atom and the imino group.

The effects of substituent groups in heteroaromatic rings show considerable regularity. Methyl (CH_3) and ethyl (C_2H_5) groups attached to ring carbon atoms usually increase the boiling point by about 20–30 °C (36–54 °F) and 50–60 °C (90–108 °F), respectively, whereas a similar attachment to a ring nitrogen atom (e.g., pyrrole → 1-methylpyrrole) significantly decreases the boiling point because of decreased ease of intermolecular association by hydrogen bonding (the active hydrogen having been replaced by a hydrocarbon group). Heterocyclic carboxylic acids and amides are all solids at room temperature. Carboxylic acids of heterocycles containing a ring nitrogen atom usually melt at higher temperatures than those containing ring oxygen or sulfur atoms, because of hydrogen bonding. Compounds containing both a ring nitrogen atom and a hydroxyl (OH) or amino (NH_2) group are usually relatively high-melting solids. Compounds containing chlorine (Cl) usually have boiling points similar to those of the corresponding ethyl-substituted compounds.

Ultraviolet, Infrared, Nuclear Magnetic Resonance and Mass Spectra

Spectroscopic studies of heterocyclic compounds, like those of other organic compounds, have became of great importance as means of identification of unknown materials, as criteria for purity, and as probes for investigating the electronic structures of molecules, thereby explaining and helping to predict their reactions. The ultraviolet spectrum of an organic compound (the pattern of its light absorption in the ultraviolet region of the spectrum) is characteristic of the π-electron system of the molecule—i.e., of the arrangement of double bonds within the structure. The ultraviolet spectra of heteroaromatic compounds show general similarity to those of benzenoid compounds (compounds with one or more benzene rings), and the effects of substituents can usually be rationalized in a similar way.

The infrared spectrum of an organic compound, with its complexity of bands, provides an excellent "fingerprint" of the compound—far more characteristic than a melting point. It also can be used to identify certain common groups, such as carbonyl (C=O) and imino, as well as various heterocyclic ring systems.

Magnetic resonance spectra are indispensable today for studies in heterocyclic chemistry. Proton resonance spectra, the most common type, yield information regarding the number of hydrogen atoms in the molecule, their chemical environment, and their relative orientation in space. Mass spectra are used to determine not only the complete molecular formula of the compound but also the molecular structure from the way the molecule fragments.

Synthesis and Modification of Heterocyclic Rings

The important methods for synthesizing heterocyclic compounds can be classified under five headings. Three are ways of forming new heterocyclic rings from precursors containing either no rings (acyclic precursors) or one fewer ring than the desired product; one is a way of obtaining a heterocyclic ring from another heterocyclic ring or from a carbocyclic ring; and one involves the modification of substituents on an existing heterocyclic ring.

In the formation of rings from acyclic precursors, the key step is frequently the formation of a carbon-heteroatom linkage (C–Z, in which Z represents an atom of nitrogen, oxygen, sulfur, or a more unusual element). The actual ring closure, or cyclization, however, may involve the formation of a carbon-carbon bond. In any case, ring formation reactions are divided into three general categories according to whether the cyclization reaction occurs primarily as a result of nucleophilic or electrophilic attack or by way of a cyclic transition state.

Nucleophilic Ring Closure

To prepare compounds containing one heteroatom, an open-chain hydrocarbon derivative containing two halogen element atoms—specifically, chlorine, bromine (Br), or iodine (I)—either as halides (in which the halogen atoms are attached directly to the hydrocarbon chain) or as acyl halides (in which the halogen atoms belong to derivatives of carboxylic acids) is reacted with the dihydro form of the heteroatom (ZH_2, or an equivalent reagent) to give nonaromatic heterocycles.

Diketones also can react with dihydro Z compounds to give heterocycles. (A ketone is an organic compound that contains a carbonyl group, the carbon atom of which is linked to two other carbon atoms belonging to hydrocarbon groups. Diketones contain two such carbonyl groups.) Diketones with the carbonyl groups separated by two carbon atoms, for example, can be cyclized to form five-membered aromatic pyrroles, furans, and thiophenes. In the case of diketones whose carbonyl groups are separated by three carbons, six-membered rings may be formed.

In each of these reactions the heteroatom Z acts as a nucleophile—an atom or a molecule that seeks a positively charged centre, such as a partly unprotected atomic nucleus. The heteroatom attacks the positively charged carbon atom produced by electron withdrawal because of the presence of the halogen atom (in the first two reactions above) or of the oxygen atom (in the last reaction).

Usually, such reactions proceed by means of intermediates in which only one of the two C–Z bonds has formed. In reactions involving halogens as halides, for instance, a compound such as $HZ-CH_2-(CH_2)_n-CH_2Br$ may form first. This fact can be applied to heterocycle synthesis in that it is frequently possible to make such intermediate compounds by other routes; these intermediates then cyclize readily to form the desired ring. One procedure for pyridine synthesis, for example, involves a condensation reaction employing an intermediate with the carbon-nitrogen bond already formed.

Heterocycles containing two adjacent nitrogen atoms, two oxygen atoms, or adjacent nitrogen and oxygen atoms also may be prepared from precursors by the use of hydrazine (N_2H_4), hydroxylamine (NH_2OH), or hydrogen peroxide (H_2O_2) in place of the dihydro Z compound. Similarly, two adjacent heteroatoms can be introduced by employing one of the reagents in the reactions with diketones.

When a compound containing two nonadjacent heteroatoms is desired, appropriate components can be put together, as in the synthesis of a pyrimidine. Ring synthesis reactions in which the heteroatom acts as a nucleophile can also employ precursors containing a ring, resulting in two-ring compounds. These reactions involve the use of ortho-disubstituted benzenes (ortho substituents being groups attached to adjacent carbon atoms in the benzene ring). The formation of quinoline and quinazoline rings is an example of this reaction type.

Electrophilic Ring Closure

Heterocyclic ring-forming reactions in which the heteroatom acts as an electrophile—an electron-seeking atom or molecule—are rare, because nitrogen, oxygen, and sulfur atoms are themselves electron-rich centres that act generally as nucleophiles. Nevertheless, electrophilic ring closure reactions are known in which a heterocyclic ring is formed by a reaction in which a carbon atom of the future ring acts as an electrophile. Usually such reactions involve ring closure onto an existing benzene ring (or other aromatic system), an electron-rich system that is generally subject to attack by electrophilic reagents. An example of ring closures of this type is the formation of quinoline from aniline and acrolein, a dehydration product of glycerol. The initial heterocyclic product of the reaction is dihydroquinoline, which must be dehydrogenated (must undergo removal of two hydrogen atoms) to give the fully aromatic product, quinoline itself.

Ring Closure by Way of Cyclic Transition States

A most important method for the synthesis of carbocyclic six-membered rings is the Diels-Alder diene reaction, named for its Nobel Prize-winning discoverers, the German chemists Otto Diels and Kurt Alder. In this reaction, illustrated below, a diene—a compound with two double bonds—reacts with a dienophile (a diene-seeking reagent), which contains a pair of carbon atoms linked by a double or triple bond. The product is a cyclohexene, a compound with a six-membered ring containing a double bond.

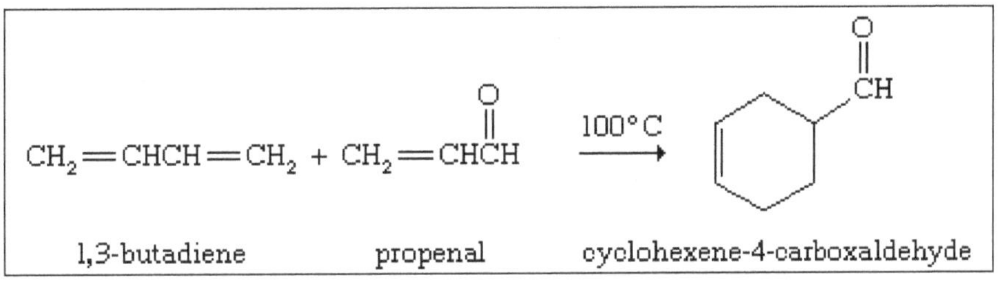

Heterocycles likewise can be synthesized by the Diels-Alder reaction, in which the dienophile contains a pair of heteroatoms such as nitrogen linked by multiple bonds. Of even greater use, however, is a related method called the Huisgen dipolar cycloaddition reaction. This reaction is an important means of preparing many types of five-membered rings, especially those containing several heteroatoms. Pyrazoles, isoxazoles, and many less-common heterocycles can be synthesized by this method.

Conversion of One Heterocyclic Ring into Another

Although there are many reactions of theoretical importance in which one heterocyclic ring is converted into another, few are of practical use. The preparation of pyridine from tetrahydrofurfuryl alcohol and ammonia and the conversion of pyrylium salts into pyridinium salts are good examples of such transformations. In addition, ring-atom rearrangement, or "shuffling," can be brought about with light in five- and six-membered heteroaromatic compounds, and ring contraction by extrusion of an atom or a group can occur under certain conditions.

Modification of an Existing Ring

Dehydrogenation of saturated or partially saturated heterocyclic rings to thermodynamically more-stable heteroaromatic compounds by heating with sulfur or by treatment with a palladium catalyst is analogous to similar reactions involving carbocyclic compounds. The hydrogenation of (addition of hydrogen to) heteroaromatic rings is, by contrast, usually more difficult, for the heteroatoms tend to poison the catalyst. Finally, the modification of substituents on heterocyclic rings is of highest importance in synthesis, and reactions by which substituents may be altered are among the most useful in heterocyclic chemistry.

Major Classes of Heterocyclic Compounds

The major classes of heterocycles containing the common heteroatoms—nitrogen, oxygen, and sulfur—are reviewed in order of increasing ring size, with compounds containing other heteroatoms. Classification by ring size is convenient because heterocyclic rings of a given size have many common features. For heterocyclic (as for carbocyclic) rings, certain broad generalizations can be made. Three- and four-membered rings, because of their small size, are geometrically strained and thus readily opened; they are also readily formed. Such heterocycles are well-known reactive intermediates. Five- and six-membered rings are readily formed and are very stable; their sizes also allow the development of aromatic character. Seven-membered rings and larger are stable but less readily formed and relatively less well investigated.

Three-membered Rings

The three-membered ring heterocycles containing single atoms of nitrogen, oxygen, and sulfur—aziridine, oxirane (or ethylene oxide), and thiirane, respectively—and their derivatives can all be prepared by nucleophilic reactions, of the type shown. Thus, aziridine is formed by heating β-aminoethyl hydrogen sulfate with a base (in this case Y is $-OSO_3H$).

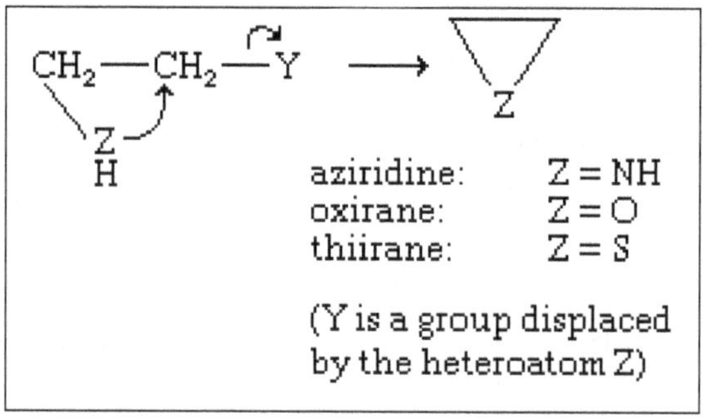

A reaction of this type is involved in the pharmacological action of nitrogen mustards, which were among the first anticancer drugs developed. Intramolecular ring closure, as in the case of the anticancer agent mechlorethamine, produces an intermediate aziridinium ion, the biologically active agent, which attacks rapidly proliferating cells such as cancer cells by inhibiting replication of their DNA (deoxyribonucleic acid). Nitrogen mustards linked to steroids also have been used as anticancer agents.

Commercially, oxirane and (to a lesser extent) aziridine are important bulk industrial chemicals. Oxirane is prepared on a large scale by the direct reaction of ethylene with oxygen.

The chemical reaction that is the most characteristic of these three-membered rings is susceptibility to attack by nucleophilic reagents to open the ring, as shown by:

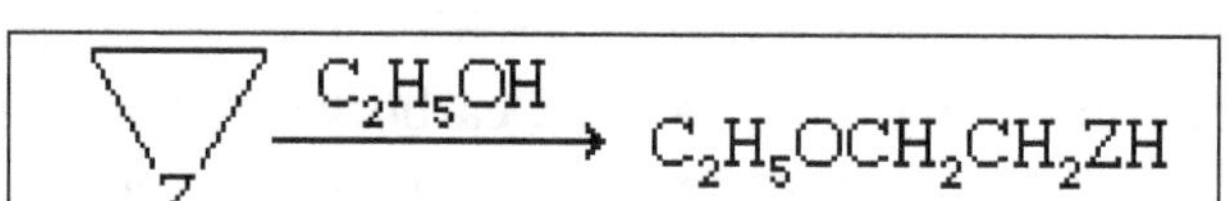

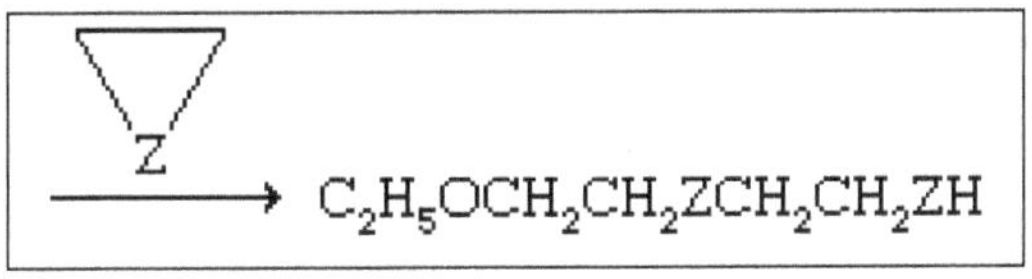

The first step shown is the reverse of the formation reaction. As indicated, a second molecule of the three-membered ring may react with the first product. Further reaction leads to long chainlike molecules (polymers) of the type $C_2H_5O(CH_2CH_2Z)_nCH_2CH_2ZH$. Polymers and copolymers of oxirane and aziridine are useful plastics.

The oxirane ring can also be opened by hydrolysis—i.e., the breaking of a bond accompanied by the addition of the elements of water (H and OH), which is similar to the first step of the ring-opening reaction illustrated above. The result of this reaction is ethylene glycol ($HOCH_2CH_2OH$), which serves as antifreeze in cooling and heating systems, in brake fluid, and as a solvent in the paints and plastics industries.

In medicine, aziridine derivatives are used for the treatment of cancer and in research as affinity probes (labeled chemicals that react selectively with biological molecules of interest) for detection and assay studies. The mitomycin family of antitumour antibiotics is the most well-known class of natural products containing the aziridine ring.

Naturally occurring compounds containing one or more oxirane rings are abundant. These molecules include insect juvenile hormones, pheromones, and products of marine organisms. Often they are biologically active and have promising therapeutic applications. For example, the fungal product (−)-ovalicin, which contains two oxirane rings, has potential for inhibiting development of solid tumours by cutting off their blood supply. An important synthetic oxirane having one ring is fosfomycin, used as an antibacterial drug, particularly in treating urinary tract infections.

Molecules containing thiirane rings are more bactericidal than those containing oxirane rings, and some thiirane derivatives have found application as tuberculostats (drugs that inhibit the growth of tuberculosis-causing bacteria), whereas thiirane 1 oxides have been reported to be insecticides, molluscicides, or herbicides.

Although aziridine, oxirane, and thiirane are saturated heterocycles, they are much more reactive than the corresponding open-chain amines, ethers, or sulfides because of the strain inherent in the three-membered ring. This behaviour reflects the comparable enhanced reactivity of the related three-membered carbocyclic compound cyclopropane.

Since 1950, five different classes of three-membered ring compounds with two heteroatoms have been discovered. They are derivatives of the parent ring systems diaziridine (containing N−N), oxaziridine (O−N), thiaziridine (S−N), dioxirane (O−O), and dithiirane (S−S). The oxathiirane system has no reported stable representatives in its class.

Except for diaziridine systems, which are used in the manufacture of plastic films and foamed plastics, as propellants and fuel stabilizers, and in the synthesis of anticancer drugs and psychopharmaceuticals, these compounds currently have limited practical importance, but they offer interesting possibilities for future applications.

Four-membered Rings

Azetidine, oxetane, and thietane—four-membered rings containing, respectively, one nitrogen, oxygen, or sulfur atom—are prepared by nucleophilic displacement reactions similar to those used to prepare the corresponding three-membered rings.

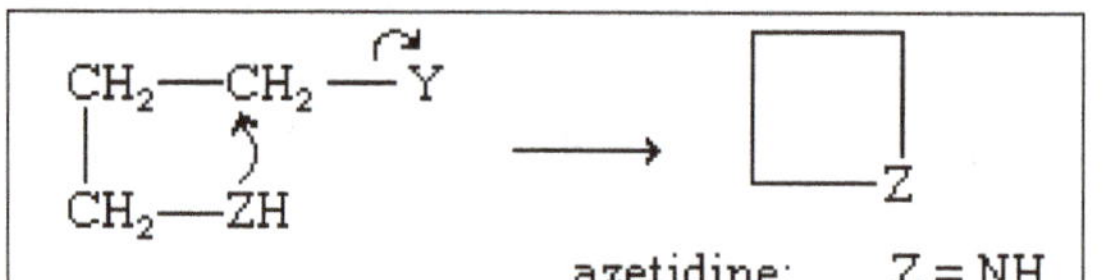

(In the reaction above, Y is usually Cl, Br, or SO_3H.) With four-membered rings, however, the reactions proceed less readily than do the analogous reactions for three-membered rings. The ring-opening reactions of four-membered heterocycles resemble qualitatively those of the corresponding three-membered rings, but they occur rather less readily.

The most important heterocycles with four-membered rings are two related series of antibiotics, the penicillins and the cephalosporins. Both series contain the azetidinone ring (the suffix -one indicating an oxygen atom linked with a double bond to a ring carbon atom). Another common name for the azetidinone ring is the β-lactam ring, which lends its name to the β-lactam antibiotics, the class to which the penicillins and cephalosporins belong. The chemistry of azetidinones was explored thoroughly during the intensive research into penicillin structure and synthesis that took place during World War II. A practical synthesis of penicillin was not achieved, however, until 1959.

Numerous oxetanes, the synthetic analogs of the antiviral natural product oxetanocin, are under investigation as antifungal, anti-inflammatory, anticancer, and antiviral agents. Oxetanones, whose ring structure is analogous to that of the azetidinones except that the heteroatom is oxygen, are widely applied in polymer manufacturing and in agriculture as herbicides, fungicides, and bactericides. The parent thietane was found in shale oil, whereas its odoriferous derivatives function as scent markers for minks, ferrets, and European polecats. Thietanes are used in the production of polymers, as bactericides and fungicides in paint, and as iron corrosion inhibitors.

Five-membered Rings with One Heteroatom

The parent aromatic compounds of this family—pyrrole, furan, and thiophene—have the structures shown.

Ergot, a fungal disease of cereal grasses, is caused by the fungus Claviceps purpurea, which produces active chemicals containing heterocyclic compounds known as indole alkaloids.

The heterocyclic dye known as indigo is a naturally occurring compound that can be obtained from plants in the genus Indigofera.

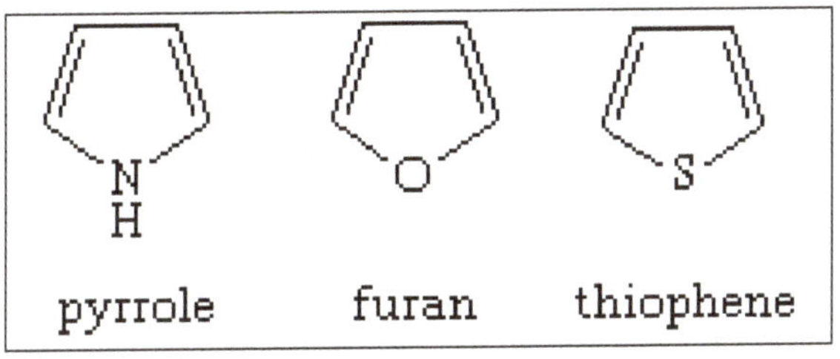

The saturated derivatives are called pyrrolidine, tetrahydrofuran, and thiophane, respectively. The bicyclic compounds made of a pyrrole, furan, or thiophene ring fused to a benzene ring are called indole (or isoindole), benzofuran, and benzothiophene, respectively.

The nitrogen heterocycle pyrrole occurs in bone oil, in which it is formed by the decomposition of proteins upon strong heating. Pyrrole rings are found in the amino acids proline and hydroxyproline, which are components of many proteins and which are present in particularly high concentrations in collagen, the structural protein of bones, tendons, ligaments, and skin.

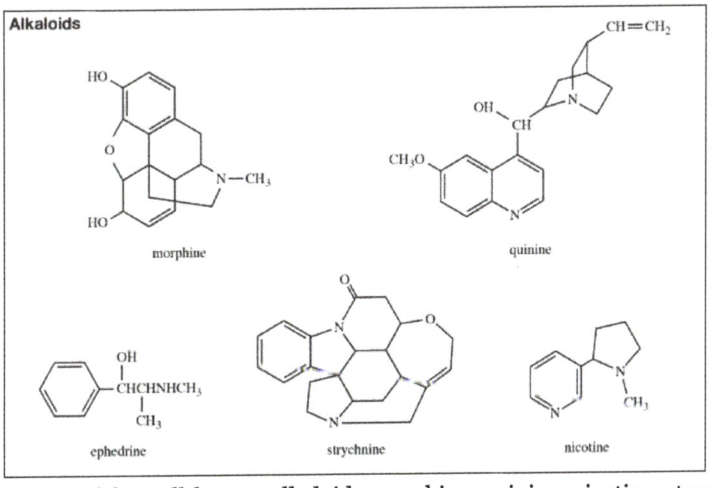

Alkaloid: Chemical structures of the well-known alkaloids morphine, quinine, nicotine, strychnine, and ephedrine.

Pyrrole derivatives are widespread in the living world. Pyrrole compounds are found among the alkaloids, a large class of alkaline organic nitrogen compounds produced primarily by plants. Nicotine is the best-known pyrrole-containing alkaloid. The heme group of the oxygen-carrying protein hemoglobin and of related compounds such as myoglobin; the chlorophylls, which are the light-gathering pigments of green plants and other photosynthetic organisms; and vitamin B_{12} are all formed from four pyrrole units joined in a larger ring system known as a porphyrin, such as that of chlorophyll b, shown below:

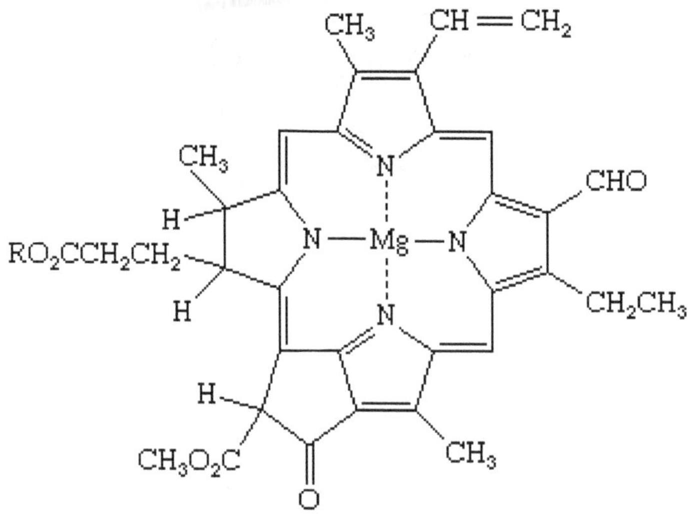

$R = \text{phytyl}, C_{20}H_{39}$
chlorophyll b
(plant pigment)

The bile pigments are formed by decomposition of the porphyrin ring and contain a chain of four pyrrole rings. Bilirubin, for example, the brownish yellow pigment that gives feces its characteristic colour, is the end product of the breakdown of heme from destroyed red blood cells.

The phthalocyanines are a group of synthetic pigments that contain four isoindole units linked together in a large ring. A typical member of the family is phthalocyanine blue (Monastral Fast Blue).

Numerous compounds produced in plants or animals contain one or more indole units in their molecular structure. The important vat dye indigo, which contains two indole units, has been used for thousands of years and was formerly obtained from plants, but it is now synthesized on a large scale.

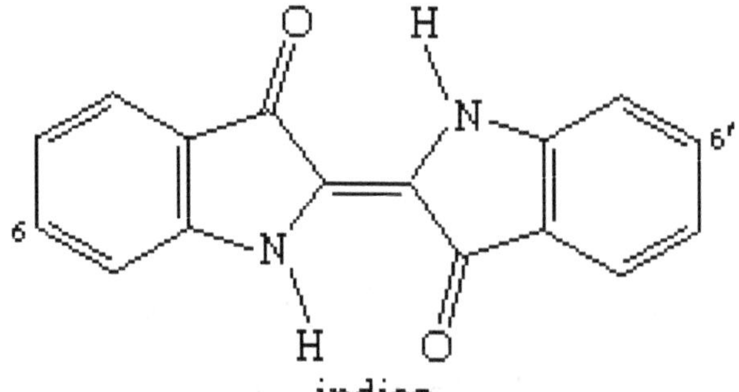

indigo

The closely related Tyrian purple, a dye obtained from species of snail and used in classical times, is 6,6′-dibromoindigo (with bromine atoms bonded to the numbered carbons in the structure above).

Tryptophan, an indole-containing essential amino acid found in most proteins, is used by the body to make several important substances, including the neurotransmitter serotonin and the B-complex vitamin niacin. Skatole, a degradation product of tryptophan that retains the indole unit, contributes much of the strong odour of mammalian feces. Indole-3-acetic acid (heteroauxin or β-indolylacetic acid) are a plant-growth regulator and the most important member of the auxin family of plant hormones. The structures of these compounds are:

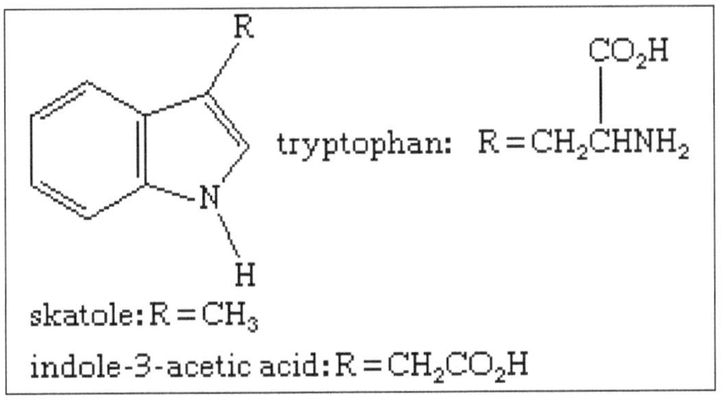

Probably the best-known indole-containing compounds are the indole alkaloids, which have been isolated from plants representing more than 30 families. The mushroom hallucinogens psilocin and psilocybin, the ergot fungus alkaloids, the drugs reserpine and yohimbine, and the poison strychnine all belong to this group.

The simplest member of the furan family of oxygen heterocycles, furan itself, is converted industrially by hydrogenation to tetrahydrofuran. Tetrahydrofuran is used as a solvent and for the production of adipic acid and hexamethylenediamine, the raw materials for nylon-6,6, the most common form of nylon. Other furan derivatives of industrial importance are maleic anhydride and phthalic anhydride, which are constituents of resins and plastics. These compounds are prepared in bulk by the oxidation of benzene and naphthalene, respectively, as shown (V_2O_5 is a vanadium catalyst).

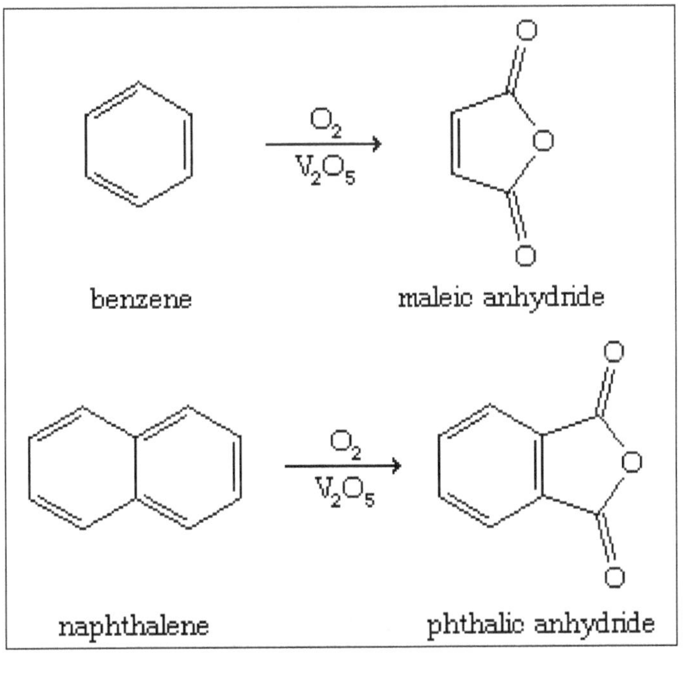

All carbohydrates, the biochemical family that includes the sugars and starches, are composed of one or more simple sugar (monosaccharide) units. These sugars are polyhydroxy aldehydes or polyhydroxy ketones that in aqueous solution exist as equilibrium mixtures of their open-chain and cyclized forms. Frequently the cyclized form of the sugar is a five-membered tetrahydrofuran ring called a furanose, as shown below for fructose, or fruit sugar, as a cyclized isomer (called a fructofuranose).

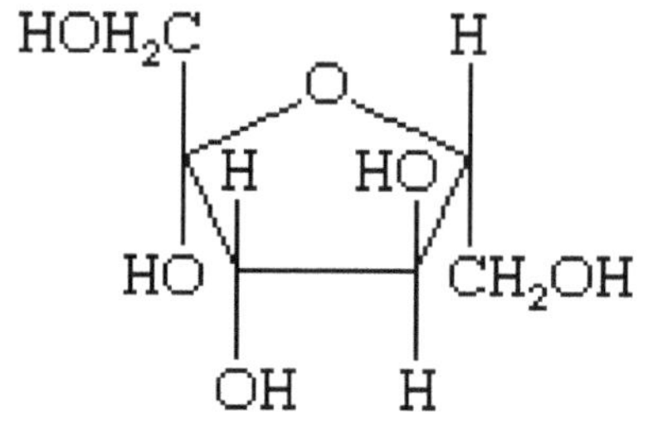

fructofuranose

Other important examples of tetrahydrofuran ring systems are the sugars ribose and deoxyribose, which are present in furanose form in, respectively, RNA and DNA, the heredity-controlling components of all living organisms.

Dehydration of certain carbohydrates yields furan derivatives. Of great commercial importance is the conversion of a carbohydrate in corncobs, oat husks, and other agricultural waste into furan-2-aldehyde, or furfural, which is used extensively as a solvent, in the manufacture of plastics, and in the preparation of other furan derivatives. Many other furan derivatives occur naturally, including vitamin C. The structures of furfural and vitamin C are:

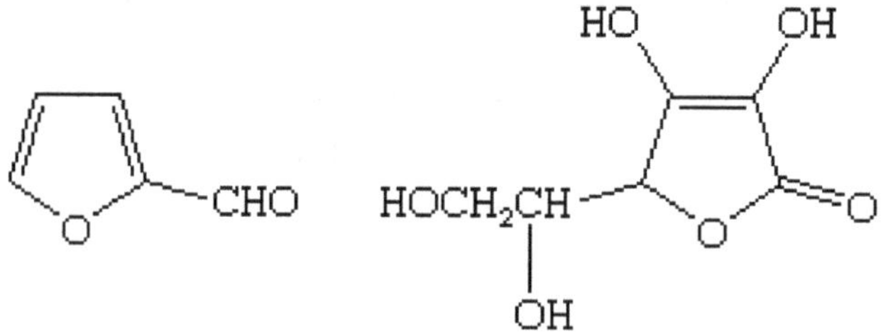

furfural

ascorbic acid (vitamin C)

The sulfur heterocycle thiophene and related compounds are found in coal tar and crude petroleum. The most important biologically occurring thiophene derivative is the B-complex vitamin biotin.

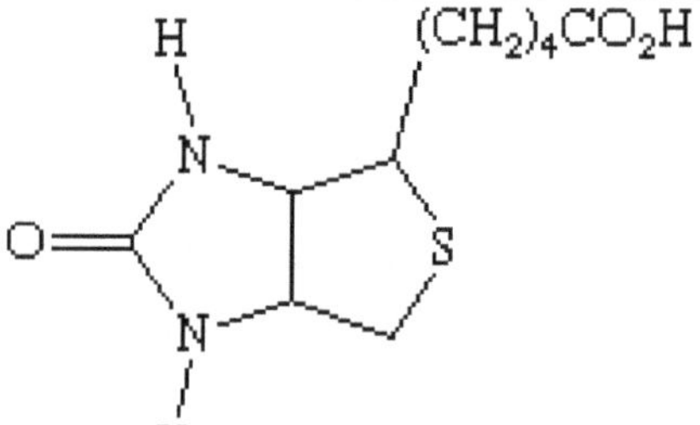

biotin (vitamin H)

Six-membered Rings with One Heteroatom

The nomenclature used for the various monocyclic nitrogen-containing six-membered ring compounds is given below. Positions on the ring are shown for pyridine, Arabic numerals being preferred to Greek letters, although both systems are used. The pyridones are aromatic compounds because of contributions to the resonance hybrid from charged resonance forms such as that shown for 4-pyridone.

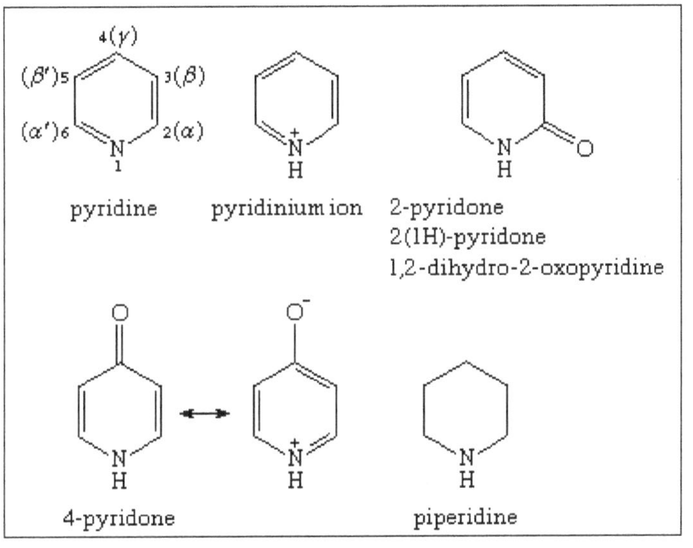

Mono-, di-, and trimethylpyridines—that is, pyridines with one, two, or three attached methyl groups, respectively—are called picolines, lutidines, and collidines, respectively, with the position of the methyl groups denoted by numbers—e.g., 2,4,6-collidine. Pyridine-2-, -3-, and -4-carboxylic acids also have widely used trivial names: picolinic, nicotinic (derived from nicotine, of which it is an oxidation product), and isonicotinic acid, respectively. Pyridine itself and the picolines, lutidines, and collidines occur in coal tar and bone oil. Pyridine derivatives are also of great biological importance. For example, nicotinic acid is more commonly known as the B-complex vitamin niacin; a nutritionally equivalent form of niacin is nicotinamide, or niacinamide. Pyridoxine is another member of the B complex, vitamin B_6. The structures of pyridoxine and nicotinamide are:

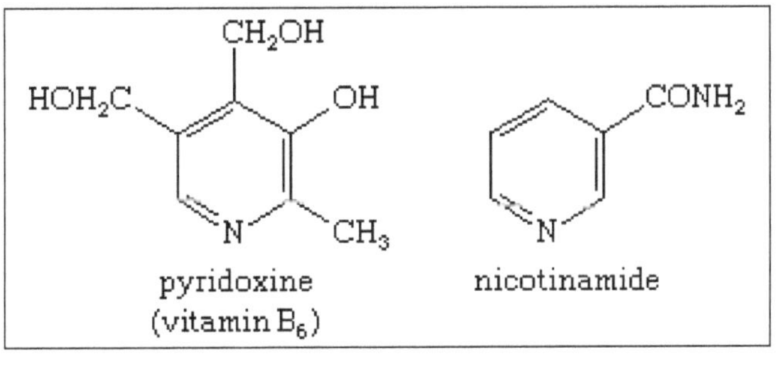

Two coenzymes involved in many important metabolic reactions in living cells, nicotinamide adenine dinucleotide (NAD, also called coenzyme I) and nicotine adenine dinucleotide phosphate (NADP, coenzyme II), are derived from nicotinamide, and the coenzyme pyridoxal phosphate (codecarboxylase) is a physiologically active form of pyridoxine. Many alkaloids contain a pyridine or piperidine ring structure, among them nicotine and piperine (one of the sharp-tasting constituents of white and black pepper, from the plant species Piper nigrum), with the structures shown.

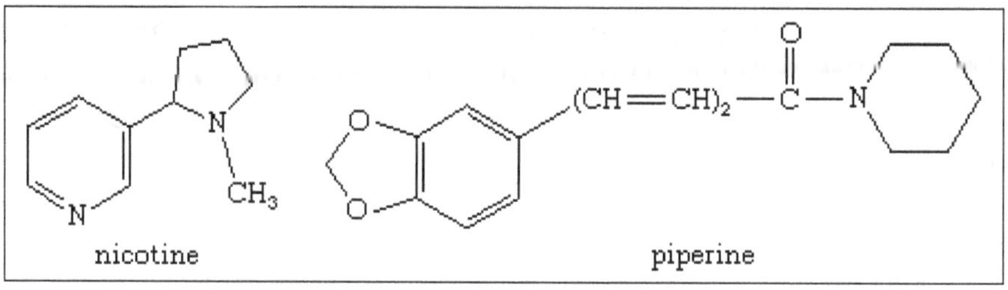

Pyridine, which once was extracted commercially from coal tar but now is prepared catalytically from tetrahydrofurfuryl alcohol and ammonia, is an important solvent and intermediate used to make other compounds. Vinylpyridines such as the structure given below, are important monomer building blocks for plastics, and fully saturated pyridine, piperidine, is used in rubber processing and as a chemical raw material.

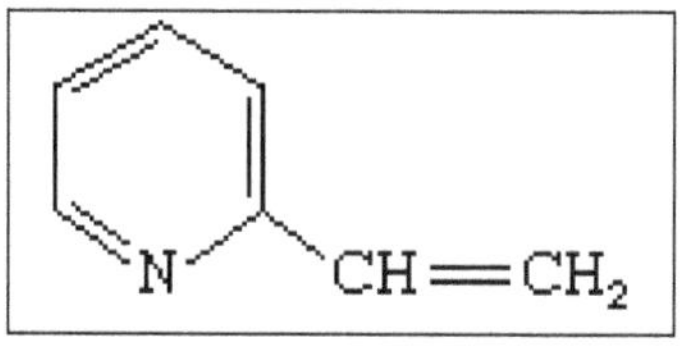

Pharmaceutically important pyridines include the tuberculostat isoniazid (isonicotinic acid hydrazide), the anti-AIDS-virus drug nevirapine, the vasodilator nicorandil, used for treating angina, the urinary-tract analgesic phenazopyridine, and the anti-inflammatory sulfa drug. 1-(1-phenylcyclohexyl) piperidine (PCP, phencyclidine) was originally used as an anesthetic, but its powerful hallucinogenic properties have led to abuse. Diquat, paraquat, clopyralid, and diflufenican are well-known pyridine derivatives used as herbicides.

The structural formulas below are two isomeric benzopyridines (upper pair) and two isomeric dibenzopyridines (lower pair), with their common names and accepted numberings. All four compounds and some of their alkyl derivatives have been obtained from coal tar. Each of them is also the parent substance of a class of alkaloids. Of these, the quinolines (e.g., quinine and other derivatives, still obtained from the Cinchona tree) and the isoquinoline (e.g., morphine) groups are particularly well-known.

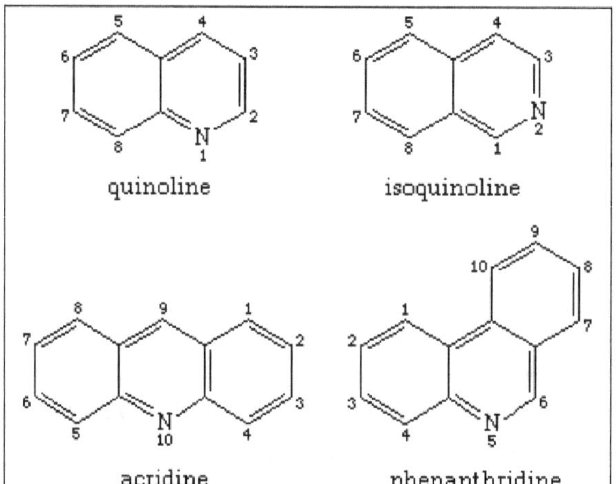

Quinoline itself can be used to manufacture nicotinic acid and other compounds such as drugs and dyes. The production of synthetic quinoline far exceeds that from coal tar. Morphine, codeine, and thebaine—all containing partially reduced isoquinoline rings—are alkaloids of the opium poppy and have been used for many centuries as hypnotics and analgesics. The semisynthetic derivative of morphine, heroin, is an even more powerful hypnotic and a highly addictive drug.

Important synthetic derivatives of the benzo- and dibenzopyridines include cyanine dyes, used as sensitizers in silver halide photographic emulsions, and quinophthalone dyes, applied in plastics, polymer textiles, paper, cosmetics, transfer-printing processes, electronic photography, and laser dyes. Other derivatives are useful bacteria-staining agents for microscopy, antiseptics such as the coal-tar dye acriflavine, antimalarial agents such as mepacrine (quinacrine) and chloroquine, antibacterial agents such as ciprofloxacin, trypanocides (drugs that destroy trypanosomes, which are parasitic protozoans responsible for Chagas disease, sleeping sickness, and other serious infectious illnesses) such as ethidium bromide, and the reagent oxine (8-hydroxyquinoline or 8-quinolinol), used in analytical chemistry.

Positively charged ions (cations) of pyrylium and thiopyrylium are the parent six-membered, aromatic, monocyclic oxygen and sulfur compounds of their respective groups.

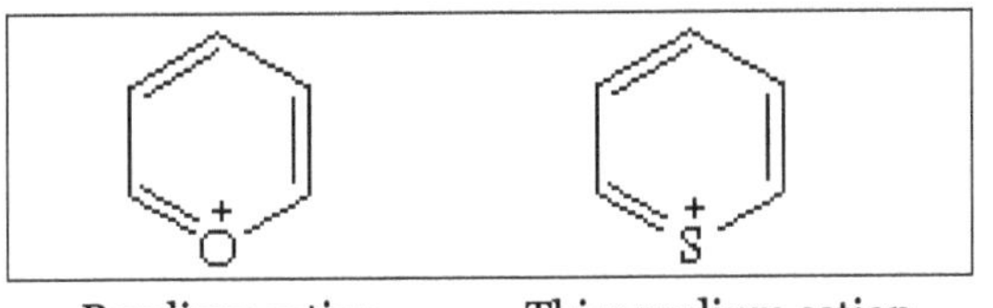

Pyrylium cation Thiopyrylium cation

An uncharged aromatic (completely conjugated) six-membered ring containing an oxygen or sulfur atom is possible only if the ring contains a carbonyl group (i.e., a ring carbon atom linked by a double bond to an oxygen atom not in the ring), as in the pyrones.

The pyrans contain extra hydrogen atoms, the position of which is indicated in structural diagrams by a number followed by an H. Certain sugars—a typical example is the monosaccharide glucose—are called pyranoses because they contain six-membered tetrahydropyran rings, with the structure:

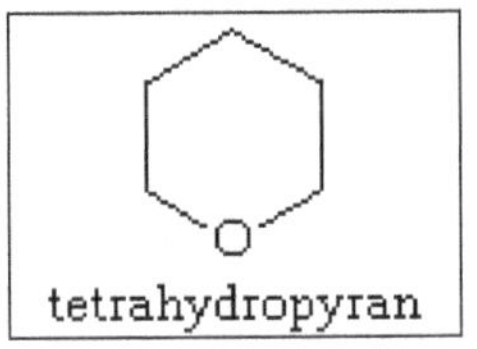

tetrahydropyran

Pyrone derivatives are present in natural products. Kojic acid, for example, is an antibiotic derived by action of certain molds on starches or sugars. The steroid bufotalin and its poisonous ester bufotoxin are obtained from the skin glands of toads.

The benzopyrylium cation is the parent of a large number of natural products. Chroman, or 3,4-dihydro-2H-1-benzopyran, is itself not found in nature, but the chroman unit is present in many natural products. Vitamin E (α-tocopherol), a substituted chroman, is found in plant oils and the leaves of green vegetables, whereas coumarin, or 2H-1-benzopyran-2-one, used in perfumes and flavourings, and its derivative dicoumarin (dicumarol, or discoumarol), a blood anticoagulant, are products of living organisms.

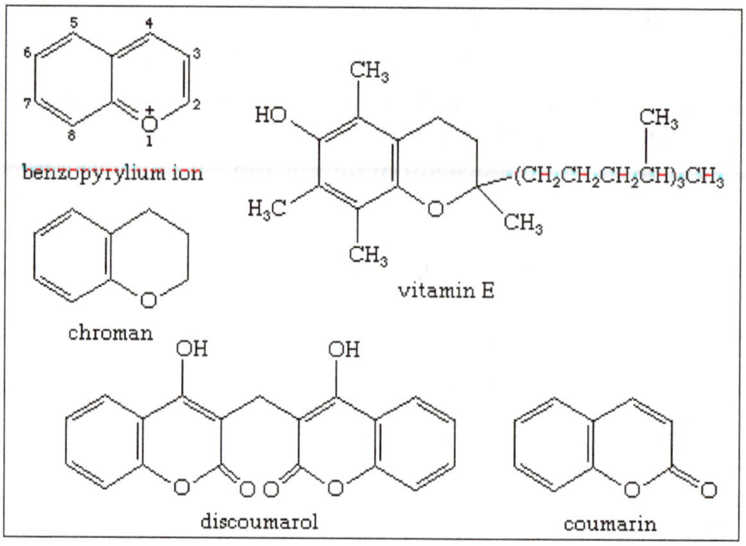

The flavylium cation is the parent of the anthocyanidines, substances that in chemical combination with sugars form the anthocyanin pigments, the common red and blue colouring matters of flowers and fruits. The colour range from yellow to reddish orange is provided by anthoxanthins, which constitute a subgroup of flavonoids. The latter are derivatives of flavone (2-phenyl-4-pyrone) in which one or more hydrogen atoms are replaced by hydroxy or methoxy groups. The flavylium ion has the structure:

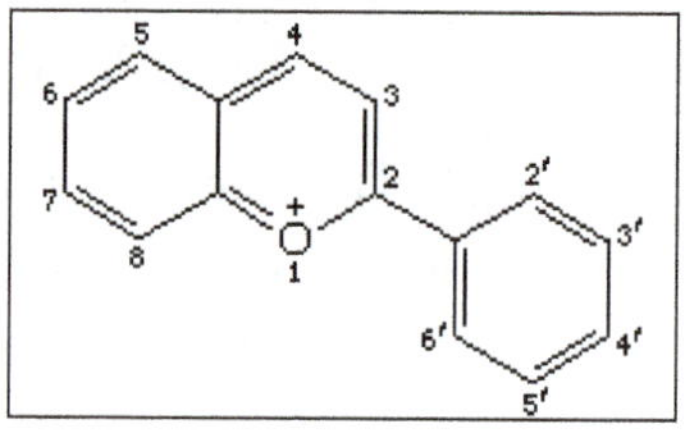

Flavylium ion

Five- and Six-membered Rings with two or more Heteroatoms

The names and numbering systems for the five-membered heteroaromatic rings with two heteroatoms are:

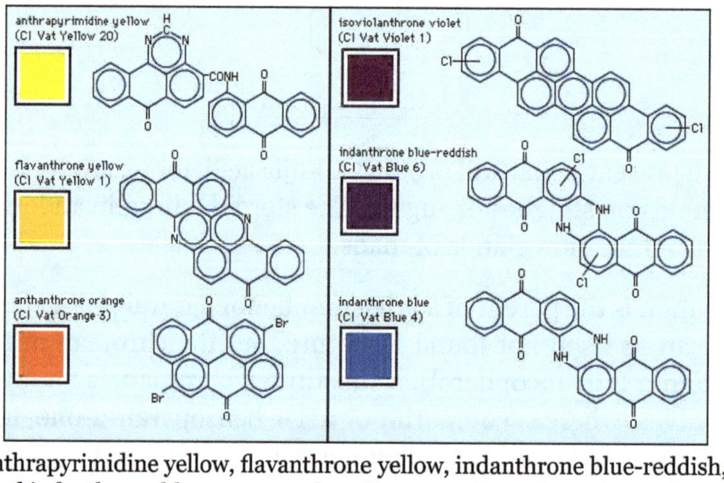

Anthrapyrimidine yellow, flavanthrone yellow, indanthrone blue-reddish, and indanthrone blue are examples of heterocyclic anthraquinone dyes.

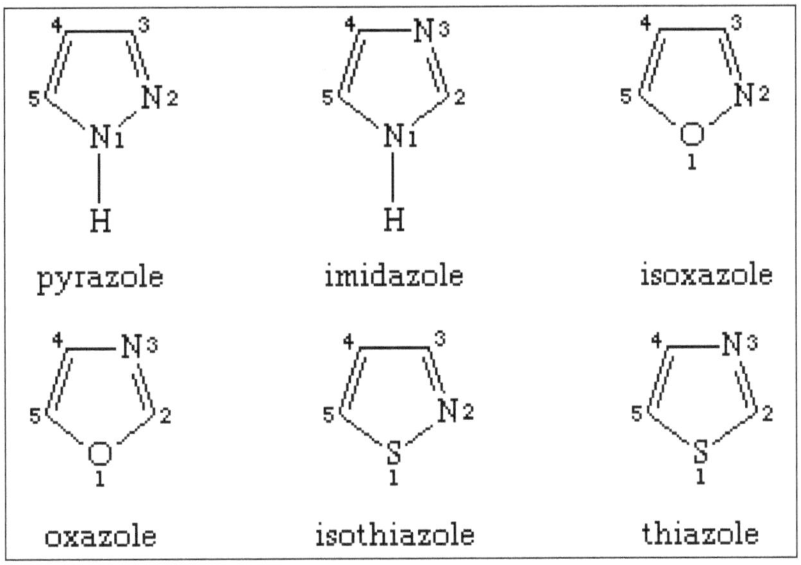

Few pyrazoles occur naturally; the compounds of this class are usually prepared by the reaction of hydrazines with 1,3-diketones. Many synthetic pyrazole compounds are of importance as dyes and medicinals. Among them are the fever-reducing analgesic aminopyrine, the anti-inflammatory drug phenylbutazone, used in treating arthritis, the yellow food colour and fibre dye tartrazine, and a series of dyes used as sensitizing agents in colour photography.

Imidazoles are most important biologically; histidine, for example, is an essential amino acid of particular importance in enzyme reactions. A breakdown product of histidine, called histamine, has a variety of functions in different organisms; in the human body it plays a crucial role in the immune response, including allergic reactions—hence the importance of antihistamine drugs. Histidine and histamine have the structures:

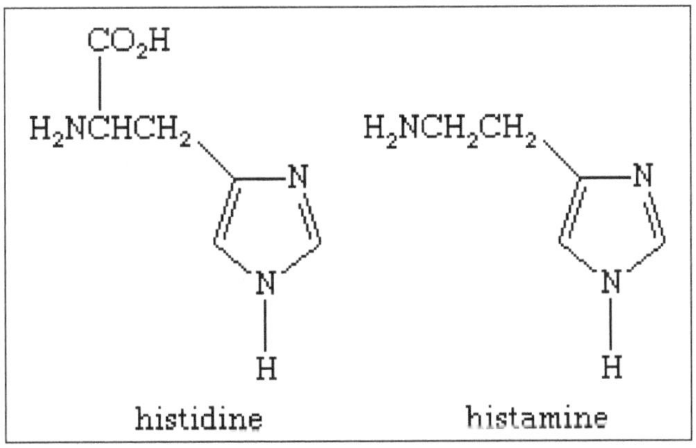

Among other naturally occurring compounds with an imidazole nucleus are hydantoin, which is found in beet sap, and allantoin, which is related to the metabolic product uric acid. Hydantoin derivatives, in particular phenytoin, are important antiepileptic drugs. The imidazole ring is also present in the B vitamin biotin.

The antibiotic cycloserine, produced by a bacterium, is one of the few naturally occurring isoxazoles. A synthetic isoxazole, hymexazol, has found practical use as soil and seed fungicide.

Thiazoles are of great biological importance. This ring system occurs in thiamin (thiamine, vitamin B_1), the bacitracin and penicillin antibiotics (from a bacterium and a mold, respectively), and in numerous synthetic drugs, dyes, and industrial chemicals. Synthetic drugs belonging to the thiazole family include the antimicrobial agents sulfathiazole and acinitrazole, the antidepressant pramipexole, and the antiasthmatic drug cinalukast. Sulfathiazole has the structure:

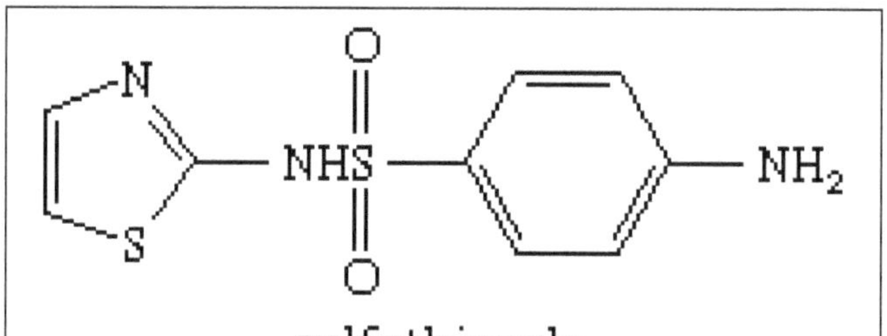

sulfathiazole

Other thiazole compounds include rhodanine, the dye rhodanine red derived from it, and the yellow dye primuline.

Most bicyclic systems derived from these five-membered rings are named systematically—that is, by use of the prefix benz- or benzo- to indicate the presence of the benzene ring. Benzimidazole, for example, is the name for the compound:

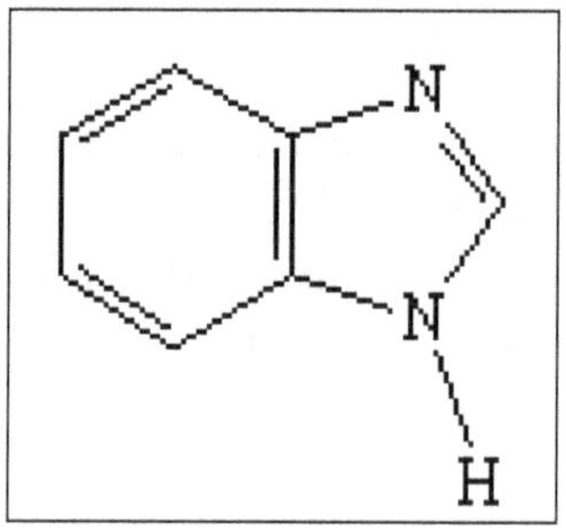

Benzimidazole

A benzimidazole unit occurs in vitamin B_{12}. Benzothiazole derivatives are used for accelerating the vulcanization of rubber (2-mercaptobenzothiazole), as herbicides (benazolin, mefenacet), and as fungicides and antihelminthic drugs (thiabendazole).

The three monocyclic diazines—six-membered ring compounds with two nitrogen heteroatoms—are named and numbered as shown.

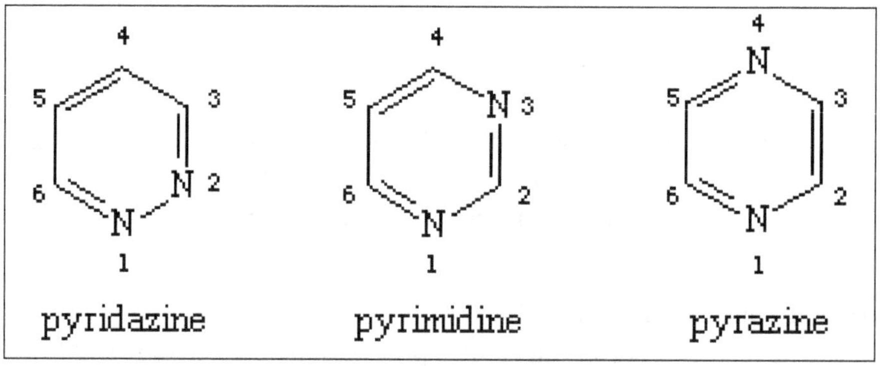

The pyridazine derivative maleic hydrazide is a herbicide, and some pyrazines occur naturally—the antibiotic aspergillic acid, for example. The structures of the aforementioned compounds are:

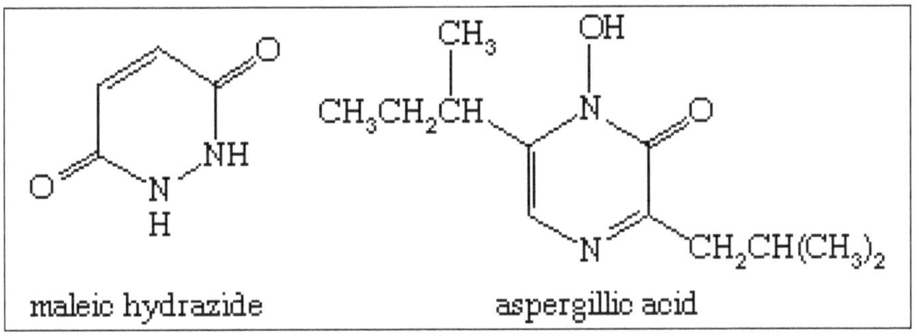

The pyrazine ring is a component of many polycyclic compounds of biological or industrial significance. Important members of the pyrazine family include pteridines, alloxazines, and phenazines.

Biologically and pharmacologically, however, the most important diazines are the pyrimidines. Uracil, thymine, and cytosine, for example, with the structures shown, are three of the five nucleotide bases that constitute the genetic code in DNA and RNA.

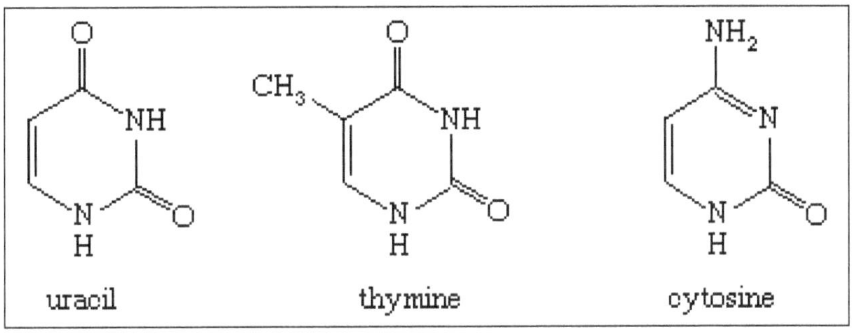

The vitamin thiamin contains a pyrimidine ring (in addition to the five-membered thiazole ring), and synthetic barbiturates such as amobarbital (amylobarbitone) are widely used drugs.

Various oxazine and thiazine derivatives are known, but monocyclic thiazines are as yet of little importance. The parent tetrahydro-1,4-oxazine, commonly called morpholine, is produced on a large scale for use as a solvent, corrosion inhibitor, and fungicide. The morpholine ring is also present in the sedative-hypnotic drug trimetozine and in some fungicides such as tridemorph and fenpropimorph. The structural formula for morpholine is:

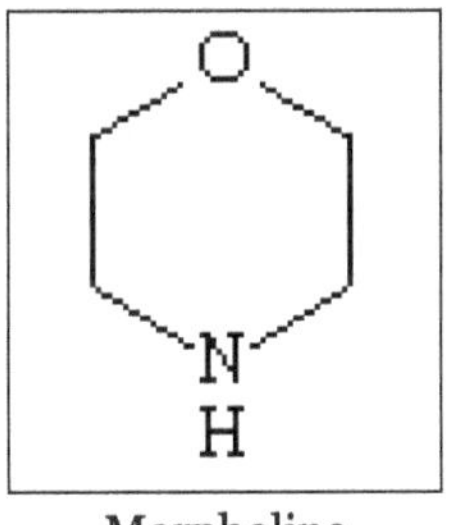

Morpholine

The benzodiazines are polycyclic compounds containing one or more benzene rings fused to a diazine ring. Many have common names—e.g., cinnoline, quinazoline, and phenazine.

Five other polycyclic systems in this general family are of significance, their names, structures, and numbering systems being:

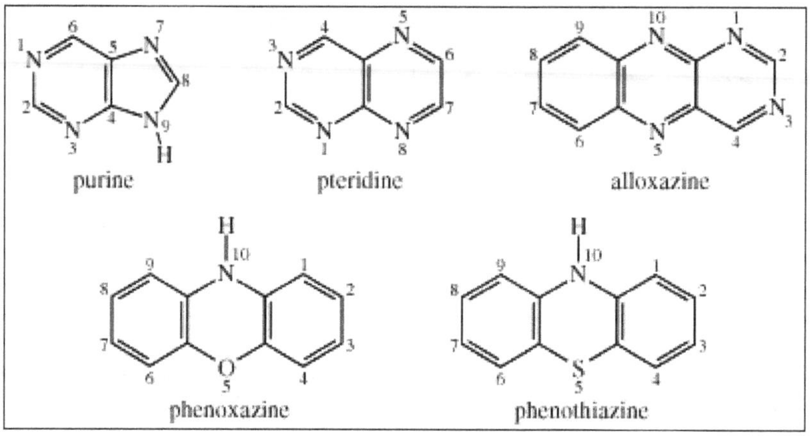

Quinoxaline alkaloids exist, and there are some phenazine natural products. Phenazine dyes are used for fabrics (for example, indanthrones and anthraquinone vat dyes; see anthraquinone dye) and for inks and printer toners (nigrosines). The first synthetic phenazine dye, mauve (aniline purple), is historically important; its structure is:

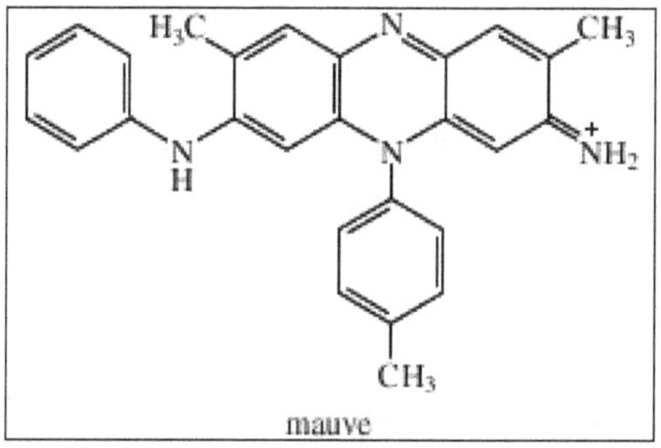

mauve

The phenoxazine system is a chromophoric (colour-imparting) part of the molecular structures of the naturally occurring actinomycin antibiotics, which are yellow-red in colour. Many polycyclic compounds containing a phenoxazine ring are used as biological stains, fabric dyes, and light-emitting materials in dye lasers (e.g., cresyl violet and nile blue).

Phenothiazine has been used as a worming agent for livestock and as an insecticide. Drugs of the phenothiazine type include the antipsychotic agents chlorpromazine and thioridazine, the long-acting antihistamine promethazine, and ethopropazine, used in treatment of parkinsonism. A large group of dyes have the phenothiazine structure, including methylene blue, a substance widely used as a biological stain and an oxidation-reduction indicator. The structure of methylene blue is:

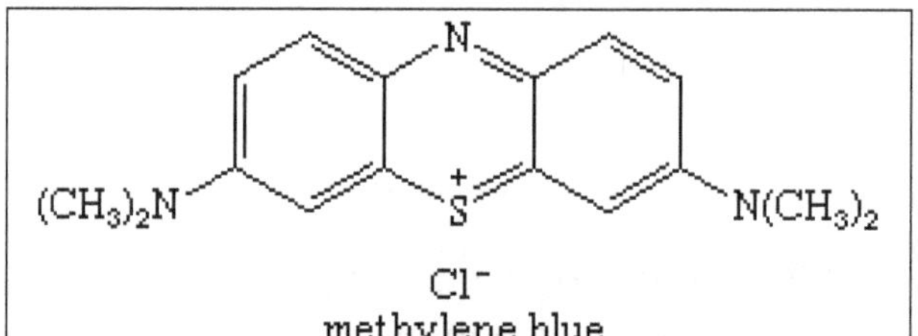

methylene blue

Biologically, the purines and pteridines are the most important polycyclic diazines. Purine itself is not common, but the purine structure is present in many natural substances. Two purine nucleotide bases, adenine and guanine, occur together with the pyrimidine bases in DNA and RNA.

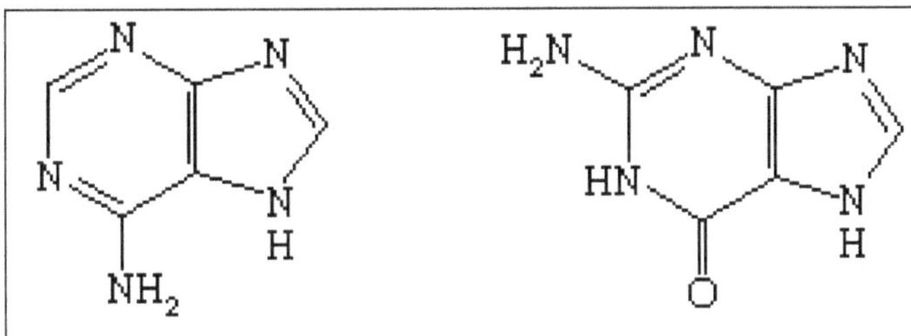

adenine guanine

Other natural purines include the alkaloids xanthine and caffeine (found in tea, coffee, and cocoa), the alkaloid theobromine (found in cocoa), and uric acid. The structures of caffeine, theobromine, and uric acid are:

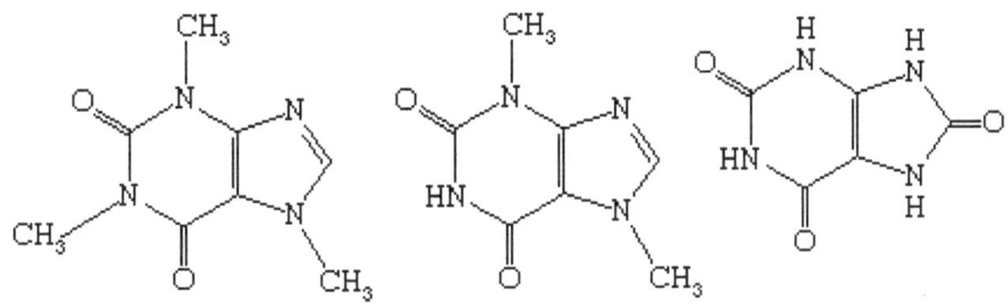

caffeine theobromine uric acid

Adenosine monophosphate, diphosphate, and triphosphate (AMP, ADP, and ATP, respectively) are important participants in energy processes in the living cell. Each of the compounds is composed of the nucleotide base adenine linked to the sugar ribose, which in turn is linked to a linear "tail" of one, two, or three phosphate groups, respectively, as shown.

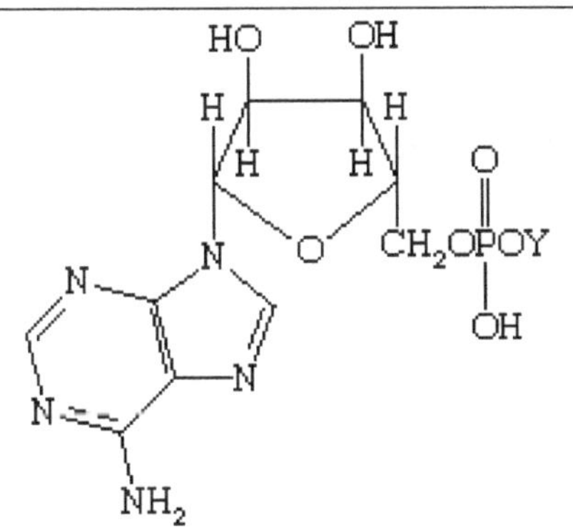

adenosine monophosphate: $Y = H$

adenosine diphosphate: $Y = PO_3H_2$

adenosine triphosphate: $Y = P_2O_6H_3$

The biological significance of pteridine compounds has become apparent since the first known members of the group were discovered as pigments of butterfly wings. One example is the yellow pigment 2-amino-4,6-pteridinedione (xanthopterin).

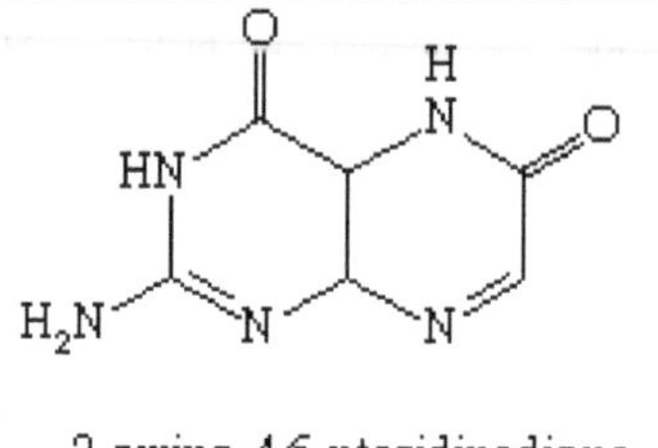

2-amino-4,6-pteridinedione

Folic acid, also a pteridine, is a B-complex vitamin and an important growth factor.

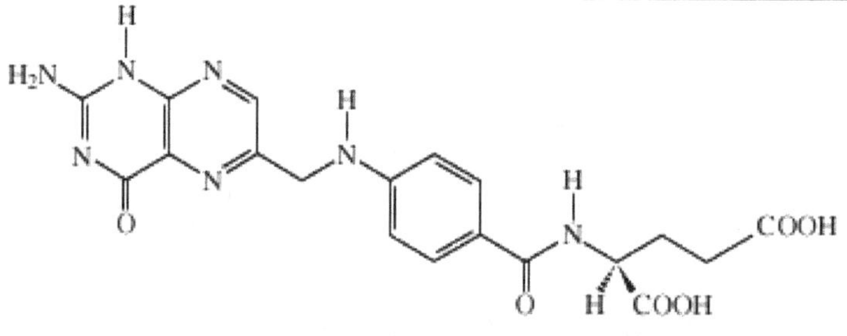

folic acid

Riboflavin, or vitamin B_2, is a derivative of alloxazine.

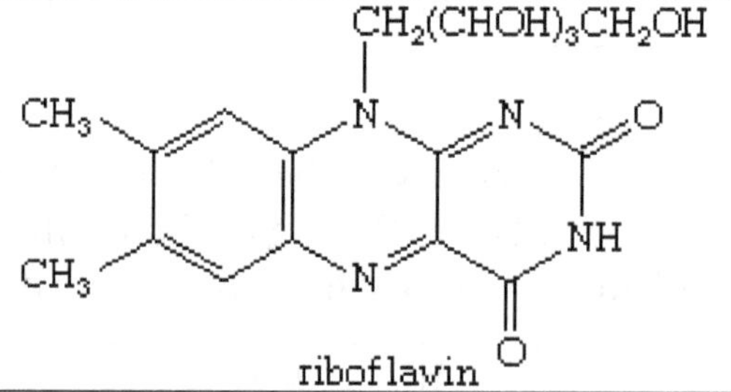

riboflavin

Rings with Seven or more Members

As the size of the ring increases, the range of compounds that can be obtained by varying the number, type, and location of the heteroatoms increases enormously. Nevertheless, the chemistry of heterocyclic compounds with rings seven-membered or larger is much less developed than that of five- and six-membered ring heterocycles, although these compounds are usually stable and some of them have found practical application. Of the seven-membered ring compounds, one-heteroatom heterocycles—azepines, oxepines, and thiepines—and their derivatives are the most comprehensively studied.

The increase in ring size constrains these compounds to be nonplanar in order to lessen the ring strain. Nonplanarity, however, affects aromaticity, so these heterocycles react as cyclic polyenes (compounds with noninteracting, alternating single and double bonds). Azepine and oxepine rings are important constituents of numerous naturally occurring alkaloids and metabolic products of marine organisms. The azepine derivative caprolactam is produced commercially in bulk for use as an intermediate in the manufacture of nylon-6 and in production of films, coatings, and synthetic leather. Seven-membered heterocycles with one or two nitrogen atoms in the ring are structural units of widely used psychopharmaceuticals such as imipramine (trade name Prazepine)—the first of the tricyclic antidepressants—and the tranquilizer diazepam (trade name Valium).

Of the larger ring heterocycles, the most important are the crown ethers, which contain one or more heterocyclic rings comprising 12 or more ring atoms and involving a number of various heteroatoms, usually nitrogen, oxygen, or sulfur. The heteroatoms are usually separated by two-carbon or three-carbon units (ethylene or propylene units, respectively). The first crown ether, dibenzo-18-crown-6, was synthesized in 1960.

The first number in a crown ether name indicates the total number of atoms involved in the macrocycle (i.e., the large ring), while the second indicates the number of heteroatoms in that ring. The remarkable feature of crown ethers, which stimulated the explosive development of their chemistry, is their ability to selectively bind the ions of metal elements (e.g., potassium and sodium) and whole organic molecules inside their cavities, the selectivity for a particular ion or molecule being directly related to the size of the macrocycle. Because of this feature, crown ethers have found wide application as ion transporters, as materials for ion-selective electrodes used in environmental testing for various metal ions, as sensitizers in photography, in medical diagnostics, and for the separation of radioactive isotopes.

Although crown ethers are not found in nature, some larger ring heterocycles that possess similar pronounced binding abilities exist as natural products. An example is the porphyrins, which are widely distributed as biological pigments—e.g., the magnesium-binding chlorophylls and the iron-binding heme groups of hemoglobin and myoglobin.

Rings with Uncommon Heteroatoms

In addition to the nitrogen, oxygen, and sulfur atoms commonly found in heterocycles, a large number of other elements form such rings—of greater or lesser stability. Such compounds are as yet of little practical importance.

Halogens, Selenium and Tellurium

Cyclic chloronium, bromonium, and iodonium ions—ions of the halogen elements chlorine, bromine, and iodine, respectively—have been prepared. Of these, only the iodine derivative has much stability.

Many heterocycles containing selenium (Se) atoms are known. Selenium shows much similarity in behaviour to sulfur; hence, selenophene, with the structure shown, resembles thiophene quite closely.

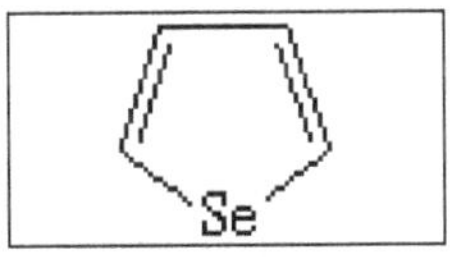

Selenophene

Carbon-selenium and selenium-selenium bonds are considerably weaker than the corresponding carbon-sulfur and sulfur-sulfur bonds; consequently, the reactions of selenium compounds tend to involve the heteroatom.

Tellurium (Te) heterocycles are rarer and even less stable than selenium heterocycles. One of the first such compounds, prepared in 1971, is dibenzotellurophene.

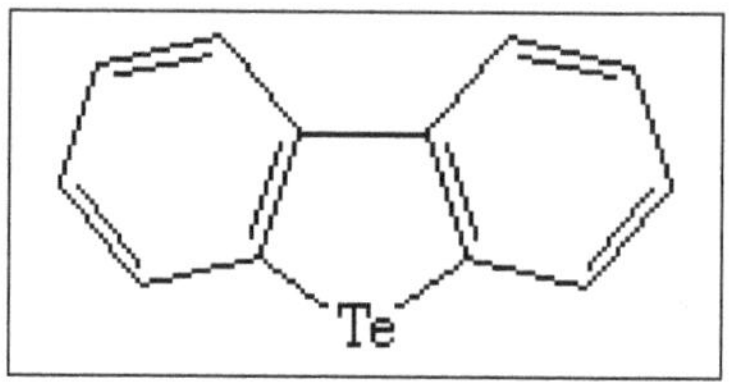

Dibenzotellurophene

Five-membered selenium heterocycles, such as selenophene-6 (featuring both selenophene and pyridine rings) and benzoselenophenes (selenium analogs of benzothiophenes;), play an important role as antioxidants in living organisms. Antioxidants reduce damage done to cells by free radicals, highly reactive molecules that are released during normal metabolic processes. Because free radicals can also be produced by exposure to ionizing radiation, these natural selenium compounds constitute potential radioprotective agents. The enhanced toxicity of the majority of selenium compounds compared with their sulfur analogs, however, significantly restricts the medical applications of these compounds.

In industry, selenium heterocycles are used almost exclusively in the preparation of "organic metals"—organic materials that have high electrical conductivity, much like metals.

Phosphorus, Arsenic, Antimony and Bismuth

Phosphorus (P), arsenic (As), antimony (Sb), and bismuth (Bi), which are all members of group Va of the periodic table of elements, form a closely related group of heterocycles. There is, however, little similarity between their properties and those of the corresponding derivatives of nitrogen, a member of the same group. Although phosphorus-containing heterocycles have long been known, the heterocyclic chemistry of arsenic, antimony, and bismuth has made significant progress only in recent years because of the decreased stability of heterocycles involving a heavy element. Parent five-membered heterocycles with arsenic, antimony, or bismuth have yet to be isolated, and six-membered bismuth heterocycles are unknown. Because of this instability, those antimony and bismuth compounds that have been synthesized have found little practical application.

Although many organophosphorus compounds are used in medicine or as insecticides and herbicides, they include few phosphorus-containing heterocycles. Two phosphorus heterocycles of practical importance are the anticancer drug cyclophosphamide and the insect chemosterilant apholate; the latter compound also contains six three-membered aziridine rings. The arsenic heterocycle 10,10′-oxybisphenoxarsine is used as a fungicidal and bactericidal additive to plastics.

Silicon, Germanium, Tin and Lead

Although silicon (Si), germanium (Ge), tin (Sn), and lead (Pb) fall in the same periodic group (IVa) as carbon, they demonstrate much lesser ability to form stable chains or double bonds.

Nevertheless, chemists know of many saturated, unsaturated, and even aromatic heterocycles that incorporate one or more atoms of these elements into their ring systems. The traditional method for preparing various such compounds is based on coupling of the halogen derivatives of the type R_2MCl_2, in which M is silicon, germanium, tin, or lead and R is halogen or a hydrocarbon chain, with organomagnesium or organolithium reagents.

All four of these elements tend to form stronger bonds with nitrogen and, especially, with oxygen than with carbon. Alternating replacement of all the carbon in carbocyclic compounds with silicon and oxygen results in silicates, which are inorganic analogs of organic heterocycles.

Boron

A large variety of heterocycles with five-, six-, or seven-membered rings containing boron (B) have been prepared and studied. Several saturated boron heterocycles were found to be more stable than their open-chain analogs, suggesting that the boron-containing cyclic structure itself favours stability. One of the best-known saturated heterocyclic boranes, widely used in organic chemistry as a reducing agent, is 9-borabicyclo nonane (9-BBN).

A boron atom and a nitrogen atom together contain the same number of electrons as two carbon atoms. Not surprisingly, a boron-nitrogen unit can replace a carbon-carbon unit in benzenoid compounds to give stable heteroaromatics. A good example is 9-aza-10-boraphenanthrene, which incorporates a boron and a nitrogen atom linked by a double bond in one of its rings.

The exhaustive substitution of all two-carbon units in a cyclohexane ring or a benzene ring with alternating boron-nitrogen units produces borazine (shown below) or borazole, respectively; the latter is often referred to as inorganic benzene.

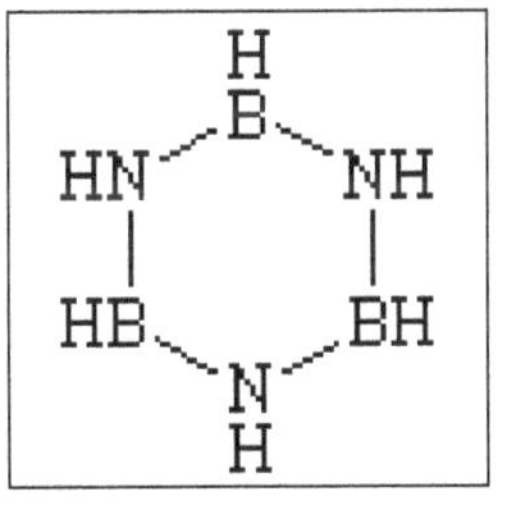

Volatile Organic Compounds

Volatile organic compounds (VOCs) are chemicals that both vaporize into air and dissolve in water. VOCs are pervasive in daily life, because they're used in industry, agriculture, transportation, and day-to-day activities around the home. Once released into groundwater, many VOCs are persistent and can migrate to drinking-water supply wells.

Formaldehyde

Formaldehyde is a simple organic compound with the formula CH_2O. It is of interest to doctors and scientists, as well as to many other industries, because of its unique chemical properties.

Formaldehyde is produced naturally by living things and some inorganic chemical reactions. It has been found via spectroscopy in interstellar space, and is produced in small quantities by our own bodies. However, like many of our body's own waste products, it is toxic to us in high concentrations. Our bodies produce just enough formaldehyde to be handled safely by the liver and kidneys.

Breathing formaldehyde fumes can cause lung and sinus irritation, sometimes severe. Long-term formaldehyde exposure is correlated to an increased risk of certain cancers. Ingesting formaldehyde (drinking it) can be fatal.

However, as humans have found more and more uses for formaldehyde, we have begun artificially producing it in large quantities. The formaldehyde industry is booming due to the chemical's vital role in industries such as car manufacture and the manufacture of building materials.

It is thought that long-term exposure to formaldehyde may increase the risk of cancer, and many regulatory agencies enforce rules about the legally allowable amount of formaldehyde in living and workspaces. The European Union has banned some products that are made with, or contain, formaldehyde.

Formaldehyde Structure

Formaldehyde consists of one carbon atom that shares a double bond with an oxygen atom. The carbon's remaining electron-sharing slots are occupied by two hydrogen atoms.

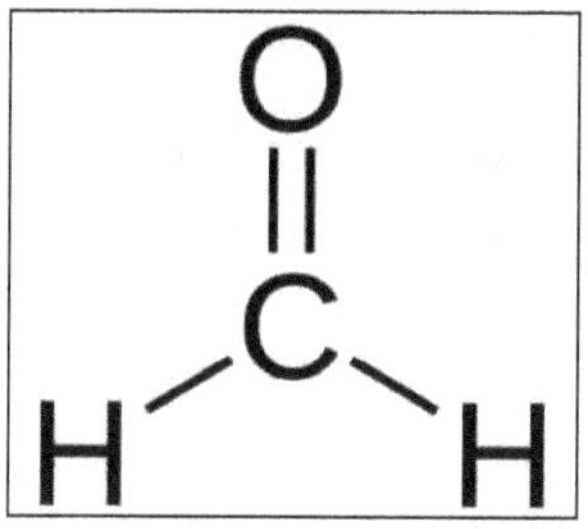

Structure of formaldehyde (methanal)

The polar nature of its carbon-oxygen bond makes this a highly reactive compound. The oxygen atom attracts electrons more strongly than the carbon atom, resulting in a partial negative charge at one end of the compound and a partial positive charge at the other. This causes it to "stick" to other polar molecules and gives it some ability to accept and donate electrons.

Its small molecular size enables it to penetrate tissues and other substances easily, while the polar nature of its carbon-oxygen bond makes it an excellent solvent. This ability to dissolve and react with many compounds is the reason it is used for many industrial and medical applications.

In its pure state at room temperature, formaldehyde is a gas. It is most useful to science and industry in solution as a liquid. This is one reason why formaldehyde fumes are common where formaldehyde is used; it is quick to come out of solution and return to its gaseous state when exposed to air.

Occurrence

Processes in the upper atmosphere contribute up to 90% of the total formaldehyde in the environment. Formaldehyde is an intermediate in the oxidation (or combustion) of methane, as well as of other carbon compounds, e.g. in forest fires, automobile exhaust, and tobacco smoke.

When produced in the atmosphere by the action of sunlight and oxygen on atmospheric methane and other hydrocarbons, it becomes part of smog. Formaldehyde has also been detected in outer space.

Formaldehyde and its adducts are ubiquitous in living organisms. It is formed in the metabolism of endogenous amino acids and is found in the bloodstream of humans and other primates at concentrations of approximately 0.1 millimolar. Experiments in which animals are exposed to an atmosphere containing isotopically labeled formaldehyde have demonstrated that even in deliberately exposed animals, the majority of formaldehyde-DNA adducts found in non-respiratory tissues are derived from endogenously produced formaldehyde.

Formaldehyde does not accumulate in the environment, because it is broken down within a few hours by sunlight or by bacteria present in soil or water. Humans metabolize formaldehyde quickly, converting it to formic acid, so it does not accumulate in the body.

Interstellar formaldehyde

Formaldehyde appears to be a useful probe in astrochemistry due to prominence of the $1_{10}\leftarrow 1_{11}$ and $2_{11}\leftarrow 2_{12}$ *K*-doublet transitions. It was the first polyatomic organic molecule detected in the interstellar medium. Since its initial detection in 1969, it has been observed in many regions of the galaxy. Because of the widespread interest in interstellar formaldehyde, it has been extensively studied, yielding new extragalactic sources. A proposed mechanism for the formation is the hydrogenation of CO ice:

$$H + CO \rightarrow HCO$$

$$HCO + H \rightarrow CH_2O$$

HCN, HNC, H_2CO, and dust have also been observed inside the comae of comets C/2012 F6 (Lemmon) and C/2012 S1 (ISON).

Synthesis and Industrial Production

Formaldehyde was first reported in 1859 by the Russian chemist Aleksandr Butlerov and was conclusively identified in 1869 by August Wilhelm von Hofmann.

1. Industry: Formaldehyde is produced industrially by the catalytic oxidation of methanol. The most common catalysts are silver metal or a mixture of an iron and molybdenum or vanadium oxides. In the commonly used formox process, methanol and oxygen react at ca. 250–400 °C in presence of iron oxide in combination with molybdenum and/or vanadium to produce formaldehyde according to the chemical equation:

$$2\ CH_3OH + O_2 \rightarrow 2\ CH_2O + 2\ H_2O$$

The silver-based catalyst usually operates at a higher temperature, about 650 °C. Two chemical reactions on it simultaneously produce formaldehyde that is shown above. The dehydrogenation reaction is given below.

$$CH_3OH \rightarrow CH_2O + H_2$$

In principle, formaldehyde could be generated by oxidation of methane, but this route is not industrially viable because the methanol is more easily oxidized than methane.

2. Organic chemistry: Formaldehyde is a building block in the synthesis of many other compounds of specialised and industrial significance. It exhibits most of the chemical properties of other aldehydes but is more reactive.

3. Self-condensation and Hydration: Formaldehyde, unlike most aldehydes, oligomerizes spontaneously. The trimer is trioxane, and the polymer is called paraformaldehyde. Many cyclic oligomers have been isolated. Similarly, formaldehyde hydrates to give the geminal diol methanediol, which condenses further to form oligomers $HO(CH_2O)_nH$. Monomeric CH_2O is rarely encountered.

4. Oxidation: It is readily oxidized by atmospheric oxygen into formic acid. For this reason, commercial formaldehyde is typically contaminated with formic acid.

5. Hydroxymethylation and Chloromethylation: Formaldehyde is a good electrophile. With good nucleophiles such as thiols, amines, and even amides, no acid catalyst is required. The resulting hydroxymethyl derivatives typically react further. Thus amines give hexahydro-1,3,5-triazines. Similarly, when combined with hydrogen sulfide, it forms trithiane.

$$3CH_2O + 3H_2S \rightarrow (CH_2S)_3 + 3H_2O$$

In the presence of acids, it participates in electrophilic aromatic substitution reactions with aromatic compounds resulting in hydroxymethylated derivatives:

$$ArH + CH_2O \rightarrow ArCH_2OH$$

When conducted in the presence of hydrogen chloride, the product is the chloromethyl compound, as described in the Blanc chloromethylation. If the arene is electron-rich, as in phenols, elaborate condensations ensue. With 4-substituted phenols one obtains calixarenes. Phenol results in polymers.

6. Base Reactions: Cannizzaro reaction in the presence of basic catalysts to produce formic acid and methanol.

Formaldehyde Uses

Because it is highly chemically reactive, formaldehyde has many uses in science and industry.

Formaldehyde uses in Biology

Formaldehyde is often used in biology to preserve tissue specimens. Formaldehyde is useful for this purpose as it kills all bacteria and fungi, and can preserve the shape of a specimen by bonding with proteins and DNA.

For the same reason, formaldehyde it is often used in embalming fluids intended to delay decay in human corpses, as well as in the preservation of animal specimens for dissection.

Formaldehyde delays, but does not permanently prevent decay of tissue. For long-term preservation of samples, scientists must use processes like plasticization, which replace the specimen's tissues with durable polymers.

In today's era of genomic analysis, scientists also sometimes prefer methods of tissue preservation that preserve nucleic acids. Formaldehyde does not preserve nucleic acids and so is not ideal for preserving tissue samples that will undergo DNA or RNA analysis. These often require special treatments to de-activate enzymes which break down DNA and RNA within cells.

Formaldehyde uses in Medicine

- Used as an antiseptic, as it kills most bacteria and fungi.
- Used in the treatment of warts and some parasites.
- Used in the production and sterilization of some vaccines.
- A formaldehyde precursor is sometimes used as an alternative to antibiotics in the treatment of urinary tract infections. The kidneys turn this precursor into formaldehyde, which is then excreted into the urinary tract instead of circulating in the blood.
- Used in some personal hygiene products to prevent bacterial growth.

Formaldehyde uses in Industry

- Used as a reactant to produce many artificial materials such as resins, plastics, and other industrial chemicals.
- Used to treat clothes to make fabrics crease-resistant.
- Used to produce materials used in numerous parts of car manufacture.
- Used in the production of plywood, carpeting, and building insulation.
- Used in the production of sanitary paper products such as napkins, paper towels, and tissues.
- Used to make chemicals used in paints and explosives.
- Used to prevent bacterial and fungal growth in animal feed for commercial farming.
- Used in the development of some types of photography film.

Formaldehyde Safety

Unfortunately, the same properties of formaldehyde that make it an excellent solvent, antiseptic, and preservative can also make it dangerous to human health. It is toxic to the human body if ingested, and can cause irritation of the skin, lungs, and sinuses which can sometimes lead to long-term problems.

The U.S. government lists formaldehyde as a "known carcinogen," meaning it has is known to increase the risk of cancer with repeated exposure. People whose jobs require regular work with formaldehyde are recommended to use safety gear to limit exposure.

Exposure to formaldehyde fumes may also make the development of asthma more likely, and can result in temporary or permanent sensitivity of the sinus passages and skin. In the United States, buildings are monitored to ensure that their air does not contain high levels of formaldehyde, which can sometimes be a risk for newly constructed buildings as formaldehyde is used in producing and finishing many building materials.

Methylene Chloride

Methylene chloride is also called dichloromethane, a colourless, volatile, practically nonflammable liquid belonging to the family of organic halogen compounds. It is extensively used as a solvent, especially in paint-stripping formulations.

Methylene chloride is commercially produced along with methyl chloride, chloroform, and carbon tetrachloride by the chlorination of methane. It boils at 40 °C (104 °F) at atmospheric pressure; it is denser than water and very slightly soluble in it. It is an effective solvent for fats, oils, greases, and many polymeric substances; its use is favoured by low toxicity, low flammability, high stability, and ease of recovery for reuse.

Uses

- Methylene chloride is predominantly used as a solvent in paint strippers and removers; as a process solvent in the manufacture of drugs, pharmaceuticals, and film coatings; as a metal cleaning and finishing solvent in electronics manufacturing; and as an agent in urethane foam blowing.
- Methylene chloride is also used as a propellant in aerosols for products such as paints, automotive products, and insect sprays.
- It is used as an extraction solvent for spice oleoresins, hops, and for the removal of caffeine from coffee. However, due to concern over residual solvent, most decaffeinators no longer use methylene chloride.
- Methylene chloride is also approved for use as a postharvest fumigant for grains and strawberries and as a degreening agent for citrus fruit.

Sources and Potential Exposure

- The principal route of human exposure to methylene chloride is inhalation of ambient air.
- Occupational and consumer exposure to methylene chloride in indoor air may be much higher, especially from spray painting or other aerosol uses. People who work in these places can breathe in the chemical or it may come in contact with the skin.
- Methylene chloride has been detected in both surface water and groundwater samples taken at hazardous waste sites and in drinking water at very low concentrations.

Assessing Personal Exposure

Several tests exist for determining exposure to methylene chloride. These tests include measurement of methylene chloride in the breath, blood, and urine. It is noted that smoking and exposure to other chemicals may affect the results of these tests.

Health Hazards

1. Acute Effects:

- Case studies of methylene chloride poisoning during paint stripping operations have demonstrated that inhalation exposure to extremely high levels can be fatal to humans.
- Acute inhalation exposure to high levels of methylene chloride in humans has resulted in effects on the central nervous system (CNS) including decreased visual, auditory, and psychomotor functions, but these effects are reversible once exposure ceases. Methylene chloride also irritates the nose and throat at high concentrations.
- Tests involving acute exposure of animals have shown methylene chloride to have moderate acute toxicity from oral and inhalation exposure.

2. Chronic Effects (Noncancer):

- The major effects from chronic inhalation exposure to methylene chloride in humans are effects on the CNS, such as headaches, dizziness, nausea, and memory loss.
- Animal studies indicate that the inhalation of methylene chloride causes effects on the liver, kidney, CNS, and cardiovascular system.
- EPA has calculated a provisional Reference Concentration (RfC) of 3 milligrams per cubic meter (mg/m^3) based on liver effects in rats. The RfC is an estimate (with uncertainty spanning perhaps an order of magnitude) of a continuous inhalation exposure to the human population (including sensitive subgroups) that is likely to be without appreciable risk of deleterious noncancer effects during a lifetime. It is not a direct estimator of risk but rather a reference point to gauge the potential effects. At exposures increasingly greater than the RfC, the potential for adverse health effects increases. Lifetime exposure above the RfC does not imply that an adverse health effect would necessarily occur.
- The Reference Dose (RfD) for methylene chloride is 0.06 milligrams per kilogram body weight per day (mg/kg/d) based on liver toxicity in rats.
- EPA has medium confidence in the RfD based on: high confidence in the study on which the RfD is based because a large number of animals of both sexes were tested in four dose groups, with a large number of controls, many effects were monitored, and a dose-related increase in severity was observed; and medium to low confidence in the database because only a few studies support the no-observed-adverse-effect level (NOAEL).

3. Reproductive/Developmental Effects:

- No studies were located regarding developmental or reproductive effects in humans from inhalation or oral exposure.
- Animal studies have demonstrated that methylene chloride crosses the placental barrier, and minor skeletal variations and lowered fetal body weights have been noted.

4. Cancer Risk:

- Several studies did not report a statistically significant increase in deaths from cancer among workers exposed to methylene chloride.
- Animal studies have shown an increase in liver and lung cancer and benign mammary gland tumors following inhalation exposure to methylene chloride.
- EPA considers methylene chloride to be a probable human carcinogen and has ranked it in EPA's Group B2.
- EPA uses mathematical models, based on animal studies, to estimate the probability of a person developing cancer from breathing air containing a specified concentration of a chemical. EPA calculated an inhalation unit risk estimate of 4.7×10^{-7} $(\mu g/m^3)^{-1}$. EPA estimates that, if an individual were to continuously breathe air containing methylene chloride at an average of 2.0 $\mu g/m^3$ (0.002 mg/m^3) over his or her entire lifetime, that person would theoretically have no more than a one-in-a-million increased chance of developing cancer as a direct result of breathing air containing this chemical. Similarly, EPA estimates that breathing air containing 20 $\mu g/m^3$ (0.02 mg/m^3) would result in not greater than a one-in-a-hundred thousand increased chance of developing cancer, and air containing 200 $\mu g/m^3$ (0.2 mg/m^3) would result in not greater than a one-in-ten thousand increased chance of developing cancer.
- EPA calculated an oral cancer slope factor of 7.5×10^{-3} $(mg/kg/d)^{-1}$.

Physical Properties

- A common synonym for methylene chloride is dichloromethane.
- Methylene chloride is a colorless liquid with a sweetish odor.
- The chemical formula for methylene chloride is CH_2Cl_2, and the molecular weight is 84.93 g/mol.
- The vapor pressure for methylene chloride is 349 mm Hg at 20 °C, and it has an octanol/water coefficient (log Kow) of 1.30.
- Methylene chloride has an odor threshold of 250 parts per million (ppm).
- Methylene chloride is slightly soluble in water and is nonflammable.

Semi-volatile Organic Compounds

Nitrobenzene

Nitrobenzene is a yellowish, oily, nitro-compound with an almond-like odor that emits toxic fumes of nitrogen oxides upon combustion. Nitrobenzene is primarily used in the manufacture of aniline, but is also used in the manufacture of lubricating oils, dyes, drugs, pesticides, and synthetic rubber. Exposure to nitrobenzene irritates the skin, eyes and respiratory tract and can result in methemoglobinemia causing fatigue, dyspnea, dizziness, disturbed vision, shortness of breath, collapse and even death. Nitrobenzene also damages the liver, spleen, kidneys and central nervous system. This substance is a possible mutagen and is reasonably anticipated to be a human carcinogen.

Carbazole

Carbazole is an organic molecule. It mainly interacts with DNA and damages it. The Carbazole ring is present in a variety of naturally occurring medicinally active substances e.g., carbazomycins and murrayafoline A. Series of carbazole derivatives including oxazinocarbazoles, isoxazolocarbazolequinone, pyrido-carbazolequinone, tetrahydrocarbazoles, benzocarbazoles, furo-carbazoles, pyridocarbazoles, pyrrolo-carbazoles, indolocarbazoles, oxazolinyl carbazoles, thienocarbazoles, imidazocarbazoles, thiazolocarbazoles, benzopyrano-carbazoles, benzofurano-carbazoles and *N*-substituted carbazoles have been synthesized and are well known for their pharmacological activities such as antioxidant, anti-inflammatory, antibacterial, antitumor, anticonvulsant, antipsychotic, antidiabetic, larvicidal properties, etc.

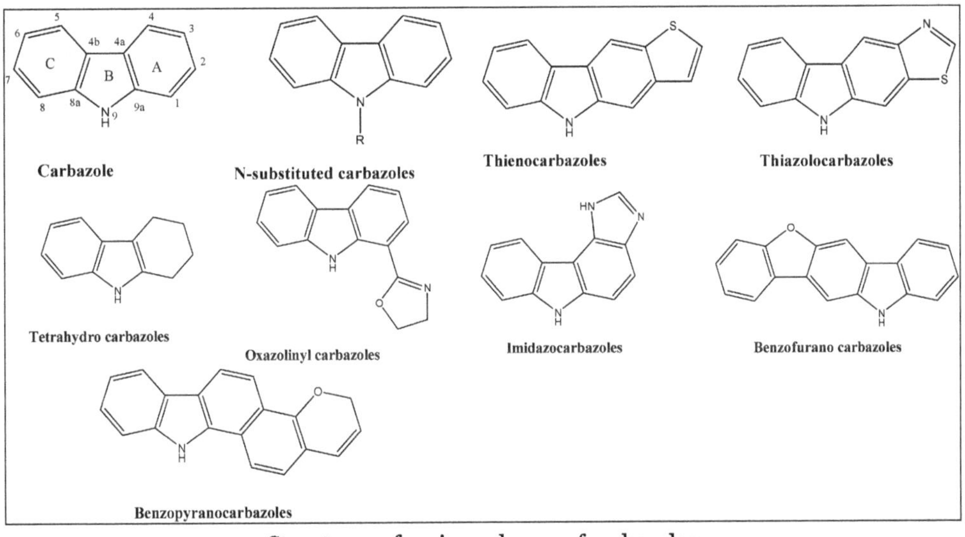

Structures of various classes of carbazoles.

Carbazole and its derivatives are an important type of nitrogen containing heterocyclic compounds that are widespread in nature. Carbazole derivatives are tricyclic heterocycles consisting of two benzene rings fused on either side of a five-membered nitrogen-containing ring. The intriguing structural features and promising pharmacological activities of these natural products have led to enormous growth in carbazole chemistry. Carbazole alkaloid derivatives are well known for their various pharmacological activities, including anti-HIV, anticancer, antibacterial and antifungal activities. Carbazole derivatives, such as *N*-vinylcarbazole and poly(vinylcarbazole) also have applications in industry and materials science as optoelectronic materials.

Pyrene

Pyrene is a polycyclic hydrocarbon (PAH) consisting of four fused benzene rings, resulting in a flat aromatic system. The chemical formula is $C_{16}H_{10}$. This yellow solid is the smallest peri-fused

PAH (one where the rings are fused through more than one face). Pyrene forms during incomplete combustion of organic compounds.

Occurrence and Properties

Pyrene was first isolated from coal tar, where it occurs up to 2% by weight. As a peri-fused PAH, pyrene is much more resonance-stabilized than its five-member-ring containing isomer fluoranthene. Therefore, it is produced in a wide range of combustion conditions. For example, automobiles produce about 1 μg/km.

Reactions

Oxidation with chromate affords perinaphthenone and then naphthalene-1,4,5,8-tetracarboxylic acid. It undergoes a series of hydrogenation reactions, and it is susceptible to halogenation, Diels-Alder additions, and nitration, all with varying degrees of selectivity. Bromination occurs at one of the 3-positions.

Photophysics

Pyrene and its derivatives are used commercially to make dyes and dye precursors, for example pyranine and naphthalene-1,4,5,8-tetracarboxylic acid. It has strong absorbance in UV-Vis in three sharp bands at 330 nm in DCM. The emission is close to the absorption, but moving at 375 nm. The morphology of the signals change with the solvent. Its derivatives are also valuable molecular probes via fluorescence spectroscopy, having a high quantum yield and lifetime (0.65 and 410 nanoseconds, respectively, in ethanol at 293 K). Pyrene was the first molecule for which excimer behavior was discovered. Such excimer appears around 450 nm.

Applications

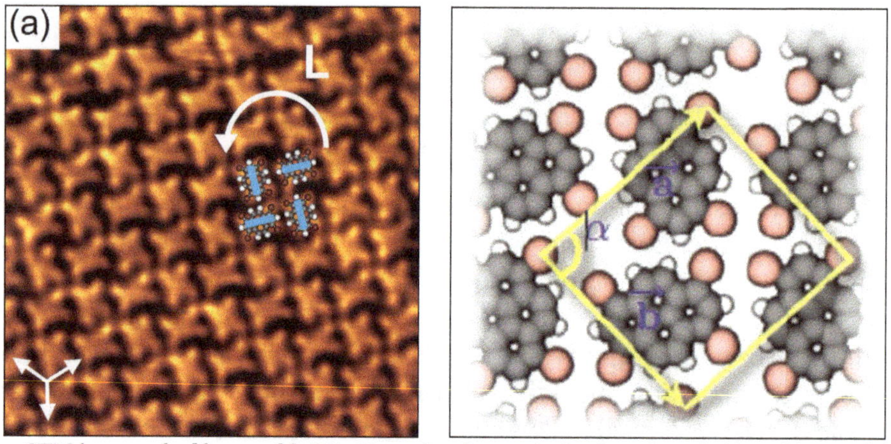

STM image of self-assembled Br_4Py molecules on Au(111) surface (left) and its model (right; pink spheres are Br atoms).

Pyrene's fluorescence emission spectrum is very sensitive to solvent polarity, so pyrene has been used as a probe to determine solvent environments. This is due to its excited state having a different, non-planar structure than the ground state. Certain emission bands are unaffected, but others vary in intensity due to the strength of interaction with a solvent.

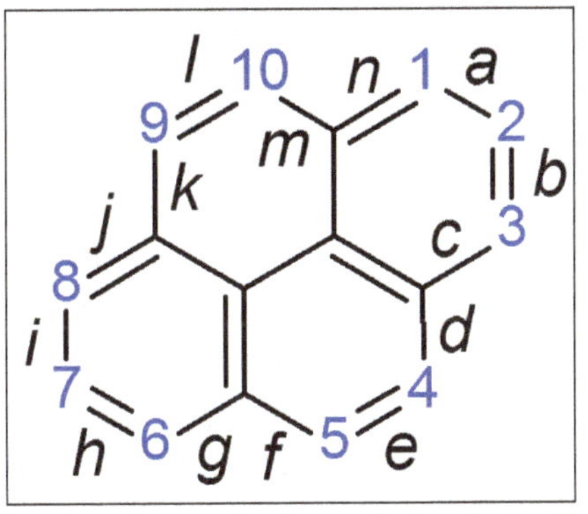

Diagram showing the numbering and ring fusion locations of pyrene according to IUPAC nomenclature of organic chemistry.

Although it is not as problematic as benzopyrene, animal studies have shown pyrene is toxic to the kidneys and liver. It is now known that pyrene affects several living functions in fish and algae. Experiments in pigs show that urinary 1-hydroxypyrene is a metabolite of pyrene, when given orally.

Pirenes are strong electron donor materials and can be combined with several materials in order to make electron donor-acceptor systems which can be used in energy conversion and light harvesting applications.

Naphthalene

Naphthalene is a white, volatile, solid polycyclic hydrocarbon with a strong mothball odor. Naphthalene is obtained from either coal tar or petroleum distillation and is primarily used to manufacture phthalic anhydride, but is also used in moth repellents.

Naphthalene is the most abundant single constituent of coal tar, a volatile product from the destructive distillation of coal, and is also formed in modern processes for the high-temperature cracking (breaking up of large molecules) of petroleum. It is commercially produced by crystallization from the intermediate fraction of condensed coal tar and from the heavier fraction of cracked petroleum. The substance crystallizes in lustrous white plates, melting at 80.1 °C (176.2 °F) and boiling at 218 °C (424 °F). It is almost insoluble in water. Naphthalene is highly volatile and has a characteristic odour; it has been used as moth repellent.

In its chemical behaviour, naphthalene shows the aromatic character associated with benzene and its simple derivatives. Its reactions are mainly reactions of substitution of hydrogen atoms by halogen atoms, nitro groups, sulfonic acid groups, and alkyl groups. Large quantities of naphthalene are converted to naphthylamines and naphthols for use as dyestuff intermediates. For many years napthalene was the principal raw material for making phthalic anhydride.

Reactivity

Like benzene, naphthalene can undergo electrophilic aromatic substitution. For many electrophilic aromatic substitution reactions, naphthalene is more reactive than benzene, and reacts under milder conditions than does benzene. For example, while both benzene and naphthalene react with chlorine in the presence of a ferric chloride or aluminum chloride catalyst, naphthalene and chlorine can react to form 1-chloronaphthalene even without a catalyst. Similarly, while both benzene and naphthalene can be alkylated using Friedel-Crafts reactions, naphthalene can also be alkylated by reaction with alkenes or alcohols, with sulfuric or phosphoric acid as the catalyst.

Mono-substitution of naphthalene has two possible isomeric products, corresponding to substitution at an alpha or beta position, respectively. Usually, the major product has the electrophile in the alpha position. The selectivity for alpha over beta substitution can be rationalized in terms of the resonance structures of the intermediate: for the alpha substitution intermediate, seven resonance structures can be drawn, of which four preserve an aromatic ring. For beta substitution, the intermediate has only six resonance structures, and only two of these are aromatic. Sulfonation, however, gives a mixture of the "alpha" product 1-naphthalenesulfonic acid and the "beta" product 2-naphthalenesulfonic acid, with the ratio dependent on reaction conditions.

Naphthalene can be hydrogenated under high pressure or with a suitable catalyst to give 1,2,3,4-tetrahydronaphthalene, a solvent sold under the trade name Tetralin. Further hydrogenation yields decahydronaphthalene or Decalin ($C_{10}H_{18}$, also known as bicyclodecane). Oxidation of naphthalene with chromate or permanganate, or catalytic oxidation with O_2 and a vanadium catalyst, gives phthalic acid.

Production

Most naphthalene is derived from coal tar. From the 1960s until the 1990s, significant amounts of naphthalene were also produced from heavy petroleum fractions during petroleum refining, but today, petroleum-derived naphthalene represents only a minor component of naphthalene production.

Although the composition of coal tar varies with the coal from which it is produced, typical coal tar is about 10% naphthalene by weight. In industrial practice, distillation of coal tar yields an oil containing about 50 percent naphthalene, along with a variety of other aromatic compounds. This oil, after being washed with aqueous sodium hydroxide to remove acidic components, chiefly various phenols, and with sulfuric acid to remove basic components, is fractionally distilled to isolate naphthalene. The crude naphthalene resulting from this process is about 95 percent naphthalene by weight. The chief impurity is the sulfur-containing aromatic compound benzothiophene. Petroleum-derived naphthalene is usually purer than that derived from coal tar. Where purer naphthalene is required, crude naphthalene can be further purified by recrystallizing it from any of a variety of solvents.

Uses

Naphthalene's most familiar use is as a household fumigant, such as in mothballs. In a sealed container of naphthalene pellets, naphthalene vapors build up to levels toxic to both the adult and larval forms of many moths that are destructive to textiles. Other fumigant uses of naphthalene include use in soil as a fumigant pesticide, and in attic spaces to repel animals. In the past, naphthalene was administered orally to kill parasitic worms in livestock.

Larger volumes of naphthalene are used as a chemical intermediate to produce other chemicals. The single largest use of naphthalene is the industrial production of phthalic anhydride, although more phthalic anhydride is made from o-xylene than from naphthalene. Other naphthalene-derived chemicals include alkyl naphthalene sulfonate surfactants, and the insecticide carbaryl. Naphthalenes substituted with combinations of strongly electron-donating functional groups, such as alcohols and amines, and strongly electron-withdrawing groups, especially sulfonic acids, are intermediates in the preparation of many synthetic dyes. The hydrogenated naphthalenes tetrahydronaphthalene (Tetralin) and decahydronaphthalene (Decalin) are used as low-volatility solvents.

Naphthalene vapor can also slow the onset of rust, and, thus, sometimes moth balls are used in places like a tool box.

Health Effects

In humans, exposure to large amounts of naphthalene may damage or destroy red blood cells. This could cause the body to have too few red blood cells until it replaces the destroyed cells. Humans, particularly children, have developed this condition after ingesting mothballs or deodorant blocks containing naphthalene. Some of the symptoms of this condition are fatigue, lack of appetite, restlessness, and pale skin. Exposure to large amounts of naphthalene may also cause nausea, vomiting, diarrhea, blood in the urine, and jaundice (yellow coloration of the skin).

Researchers with the U.S. National Toxicology Program exposed male and female rats and mice to naphthalene vapors on weekdays for two years. They found that female mice exhibited some evidence of carcinogenic activity, based on increased incidence of alveolar and bronchiolar adenomas of the lung; but male mice exhibited no evidence of carcinogenic activity. In both male and female mice, naphthalene led to greater incidences and severity of chronic inflammation and abnormalities in the tissues lining the nose and lungs. Regarding rats, they found clear evidence of carcinogenic activity of naphthalene in male and female rats based on increased incidences of respiratory epithelial adenoma and olfactory epithelial neuroblastoma of the nose. Exposure to naphthalene caused significant increases in the incidences of nonneoplastic lesions of the nose in both male and female rats.

Over 400 million people have an inherited condition called glucose-6-phosphate dehydrogenase deficiency (G6PD deficiency). For these people, exposure to naphthalene is harmful and may cause hemolytic anemia, which causes their erythrocytes to break down.

The International Agency for Research on Cancer (IARC) classifies naphthalene as possibly carcinogenic to humans [Group 2B]. It also points out that acute exposure causes cataracts in humans, rats, rabbits, and mice. Furthermore, hemolytic anemia can occur in children and infants after oral or inhalation exposure, or after maternal exposure during pregnancy.

Organic Compounds Essential to Human Functioning

Organic compounds typically consist of groups of carbon atoms covalently bonded to hydrogen, usually oxygen, and often other elements as well. Created by living things, they are found throughout the world, in soils and seas, commercial products, and every cell of the human body. The four types most important to human structure and function are carbohydrates, lipids, proteins, and nucleotides. Before exploring these compounds, you need to first understand the chemistry of carbon.

The Chemistry of Carbon

What makes organic compounds ubiquitous is the chemistry of their carbon core. Recall that carbon atoms have four electrons in their valence shell, and that the octet rule dictates that atoms tend to react in such a way as to complete their valence shell with eight electrons. Carbon atoms do not complete their valence shells by donating or accepting four electrons. Instead, they readily share electrons via covalent bonds.

Commonly, carbon atoms share with other carbon atoms, often forming a long carbon chain referred to as a carbon skeleton. When they do share, however, they do not share all their electrons exclusively with each other. Rather, carbon atoms tend to share electrons with a variety of other elements, one of which is always hydrogen. Carbon and hydrogen groupings are called hydrocarbons.

Many combinations are possible to fill carbon's four "vacancies." Carbon may share electrons with oxygen or nitrogen or other atoms in a particular region of an organic compound. Moreover, the atoms to which carbon atoms bond may also be part of a functional group. A functional group is a group of atoms linked by strong covalent bonds and tending to function in chemical reactions as a single unit. You can think of functional groups as tightly knit "cliques" whose members are unlikely to be parted. Five functional groups are important in human physiology; these are the hydroxyl, carboxyl, amino, methyl and phosphate groups.

Functional Groups Important in Human Physiology		
Functional group	Structural formula	Importance
Hydroxyl	—O—H	Hydroxyl groups are polar. They are components of all four types of organic compounds. They are involved in dehydration synthesis and hydrolysis reactions.
Carboxyl	O—C—OH	Carboxyl groups are found within fatty acids, amino acids, and many other acids.
Amino	$—N—H_2$	Amino groups are found within amino acids, the building blocks of proteins.
Methyl	$—C—H_3$	Methyl groups are found within amino acids.
Phosphate	$—P—O_4^{2-}$	Phosphate groups are found within phospholipids and nucleotides.

Carbon's affinity for covalent bonding means that many distinct and relatively stable organic molecules nevertheless readily form larger, more complex molecules. Any large molecule is referred to as macromolecule (macro- = "large"). However, some macromolecules are made up of several "copies" of single units called monomer (mono- = "one"; -mer = "part"). Like beads in a long necklace, these monomers link by covalent bonds to form long polymers (poly- = "many"). There are many examples of monomers and polymers among the organic compounds.

Monomers form polymers by engaging in dehydration synthesis. This reaction results in the release of a molecule of water. Each monomer contributes: One gives up a hydrogen atom and the other gives up a hydroxyl group. Polymers are split into monomers by hydrolysis (-lysis = "rupture"). The bonds between their monomers are broken, via the donation of a molecule of water, which contributes a hydrogen atom to one monomer and a hydroxyl group to the other.

Carbohydrates

The term carbohydrate means "hydrated carbon." Recall that the root hydro- indicates water. A carbohydrate is a molecule composed of carbon, hydrogen, and oxygen; in most carbohydrates, hydrogen and oxygen are found in the same two-to-one relative proportions they have in water. In fact, the chemical formula for a "generic" molecule of carbohydrate is $(CH_2O)_n$.

Carbohydrates are referred to as saccharides, a word meaning "sugars." Three forms are important in the body. Monosaccharides are the monomers of carbohydrates. Disaccharides (di- = "two") are made up of two monomers. Polysaccharides are the polymers, and can consist of hundreds to thousands of monomers.

Monosaccharides

A monosaccharide is a monomer of carbohydrates. Five monosaccharides are important in the body. Three of these are the hexose sugars, so called because they each contain six atoms of carbon. These are glucose, fructose, and galactose. The remaining monosaccharides are the two pentose sugars, each of which contains five atoms of carbon. They are ribose and deoxyribose.

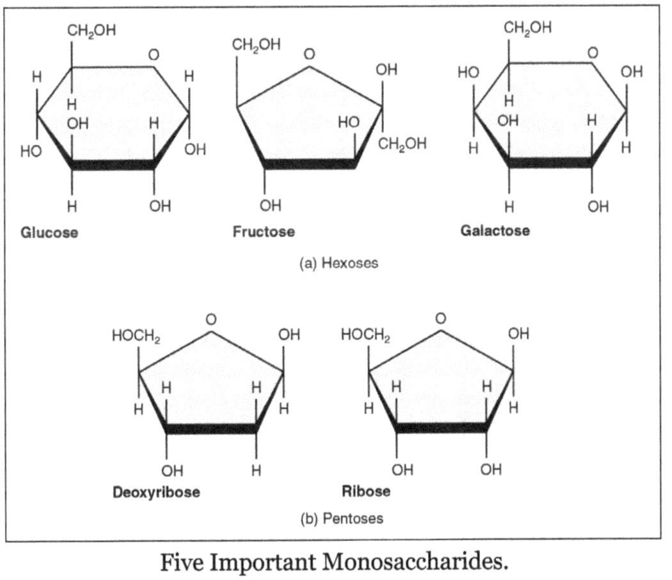

Five Important Monosaccharides.

Disaccharides

A disaccharide is a pair of monosaccharides. Disaccharides are formed via dehydration synthesis, and the bond linking them is referred to as a glycosidic bond (glyco- = "sugar"). Three disaccharides are important to humans. These are sucrose, commonly referred to as table sugar; lactose, or milk sugar; and maltose, or malt sugar. As you can tell from their common names, you consume these in your diet; however, your body cannot use them directly. Instead, in the digestive tract, they are split into their component monosaccharides via hydrolysis.

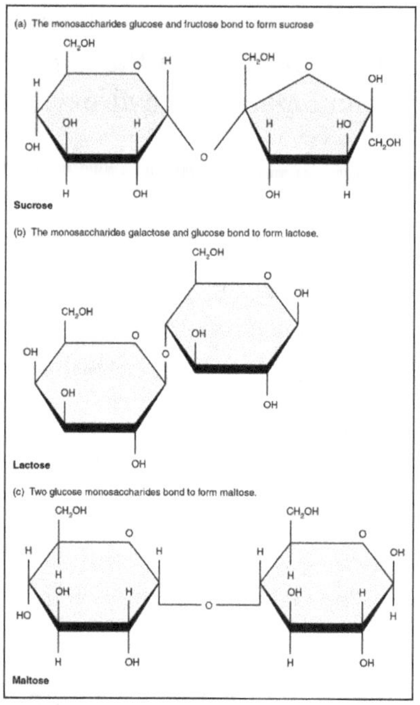

Three Important Disaccharides. All three important disaccharides form by dehydration synthesis.

Polysaccharides

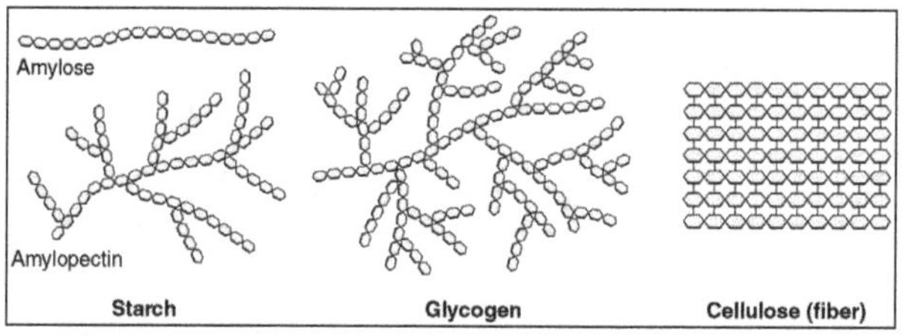

Three Important Polysaccharides. Three important polysaccharides are starches, glycogen, and fiber.

Polysaccharides can contain a few to a thousand or more monosaccharides. Three are important to the body:

- Starches are polymers of glucose. They occur in long chains called amylose or branched chains called amylopectin, both of which are stored in plant-based foods and are relatively easy to digest.
- Glycogen is also a polymer of glucose, but it is stored in the tissues of animals, especially in the muscles and liver. It is not considered a dietary carbohydrate because very little glycogen remains in animal tissues after slaughter; however, the human body stores excess glucose as glycogen, again, in the muscles and liver.
- Cellulose, a polysaccharide that is the primary component of the cell wall of green plants, is the component of plant food referred to as "fiber". In humans, cellulose/fiber is not digestible; however, dietary fiber has many health benefits. It helps you feel full so you eat less, it promotes a healthy digestive tract, and a diet high in fiber is thought to reduce the risk of heart disease and possibly some forms of cancer.

Functions of Carbohydrates

The body obtains carbohydrates from plant-based foods. Grains, fruits, and legumes and other vegetables provide most of the carbohydrate in the human diet, although lactose is found in dairy products.

Although most body cells can break down other organic compounds for fuel, all body cells can use glucose. Moreover, nerve cells (neurons) in the brain, spinal cord, and through the peripheral nervous system, as well as red blood cells, can use only glucose for fuel. In the breakdown of glucose for energy, molecules of adenosine triphosphate, better known as ATP, are produced. Adenosine triphosphate (ATP) is composed of a ribose sugar, an adenine base, and three phosphate groups. ATP releases free energy when its phosphate bonds are broken, and thus supplies ready energy to the cell. More ATP is produced in the presence of oxygen (O_2) than in pathways that do not use oxygen. The overall reaction for the conversion of the energy in glucose to energy stored in ATP can be written:

$$C_6H_{12}O_6 + 6\,O_2 \rightarrow 6\,CO_2 + 6\,H_2O + ATP$$

In addition to being a critical fuel source, carbohydrates are present in very small amounts in cells' structure. For instance, some carbohydrate molecules bind with proteins to produce glycoproteins, and others combine with lipids to produce glycolipids, both of which are found in the membrane that encloses the contents of body cells.

Lipids

A lipid is one of a highly diverse group of compounds made up mostly of hydrocarbons. The few oxygen atoms they contain are often at the periphery of the molecule. Their nonpolar hydrocarbons make all lipids hydrophobic. In water, lipids do not form a true solution, but they may form an emulsion, which is the term for a mixture of solutions that do not mix well.

Triglycerides

A triglyceride is one of the most common dietary lipid groups, and the type found most abundantly in body tissues. This compound, which is commonly referred to as a fat, is formed from the synthesis of two types of molecules:

- A glycerol backbone at the core of triglycerides, consists of three carbon atoms.
- Three fatty acids, long chains of hydrocarbons with a carboxyl group and a methyl group at opposite ends, extend from each of the carbons of the glycerol.

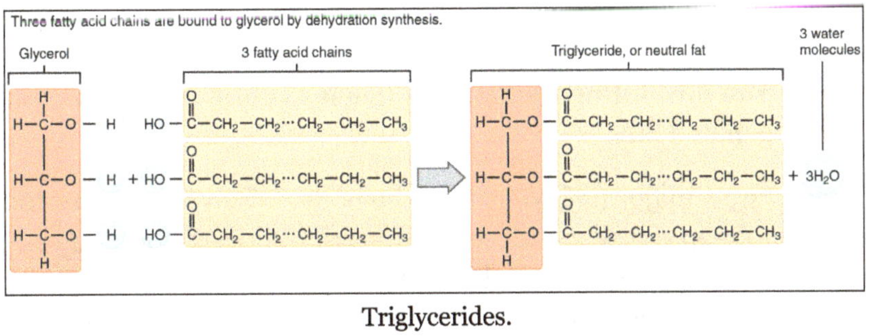

Triglycerides.

Triglycerides are composed of glycerol attached to three fatty acids via dehydration synthesis. Notice that glycerol gives up a hydrogen atom, and the carboxyl groups on the fatty acids each give up a hydroxyl group.

Triglycerides form via dehydration synthesis. Glycerol gives up hydrogen atoms from its hydroxyl groups at each bond, and the carboxyl group on each fatty acid chain gives up a hydroxyl group. A total of three water molecules are thereby released.

Fatty acid chains that have no double carbon bonds anywhere along their length and therefore contain the maximum number of hydrogen atoms are called saturated fatty acids. These straight, rigid chains pack tightly together and are solid or semi-solid at room temperature. Butter and lard are examples, as is the fat found on a steak or in your own body. In contrast, fatty acids with one double carbon bond are kinked at that bond. These monounsaturated fatty acids are therefore unable to pack together tightly, and are liquid at room temperature. Polyunsaturated fatty acids contain two or more double carbon bonds, and are also liquid at room temperature. Plant oils such as olive oil typically contain both mono- and polyunsaturated fatty acids.

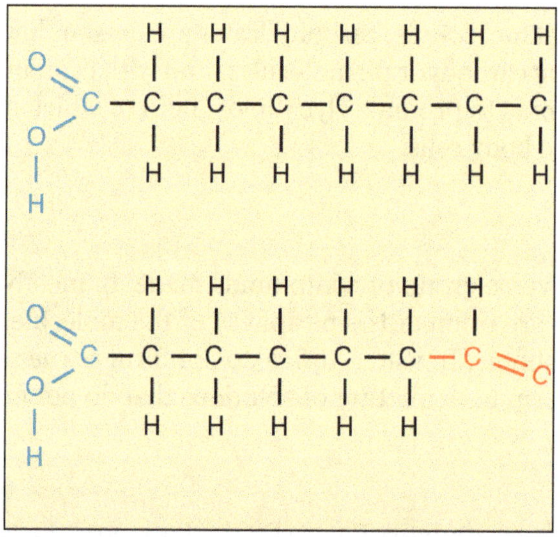

Fatty Acid Shapes. The level of saturation of a fatty acid affects its shape.
(a) Saturated fatty acid chains are straight. (b) Unsaturated fatty acid chains are kinked.

Whereas a diet high in saturated fatty acids increases the risk of heart disease, a diet high in unsaturated fatty acids is thought to reduce the risk. This is especially true for the omega-3 unsaturated fatty acids found in cold-water fish such as salmon. These fatty acids have their first double carbon bond at the third hydrocarbon from the methyl group (referred to as the omega end of the molecule).

Finally, trans fatty acids found in some processed foods, including some stick and tub margarines, are thought to be even more harmful to the heart and blood vessels than saturated fatty acids. Trans fats are created from unsaturated fatty acids (such as corn oil) when chemically treated to produce partially hydrogenated fats.

As a group, triglycerides are a major fuel source for the body. When you are resting or asleep, a majority of the energy used to keep you alive is derived from triglycerides stored in your fat (adipose) tissues. Triglycerides also fuel long, slow physical activity such as gardening or hiking, and contribute a modest percentage of energy for vigorous physical activity. Dietary fat also assists the absorption and transport of the nonpolar fat-soluble vitamins A, D, E, and K. Additionally, stored body fat protects and cushions the body's bones and internal organs, and acts as insulation to retain body heat.

Fatty acids are also components of glycolipids, which are sugar-fat compounds found in the cell membrane. Lipoproteins are compounds in which the hydrophobic triglycerides are packaged in protein envelopes for transport in body fluids.

Phospholipids

As its name suggests, a phospholipid is a bond between the glycerol component of a lipid and a phosphorous molecule. In fact, phospholipids are similar in structure to triglycerides. However, instead of having three fatty acids, a phospholipid is generated from a diglyceride, a glycerol with just two fatty acid chains. The third binding site on the glycerol is taken up by the phosphate group, which in turn is attached to a polar "head" region of the molecule. Recall that triglycerides are nonpolar and hydrophobic. This still holds for the fatty acid portion of a phospholipid compound. However, the head of a phospholipid contains charges on the phosphate groups, as well as on the nitrogen atom. These charges make the phospholipid head hydrophilic. Therefore, phospholipids are said to have hydrophobic tails, containing the neutral fatty acids, and hydrophilic heads, containing the charged phosphate groups and nitrogen atom.

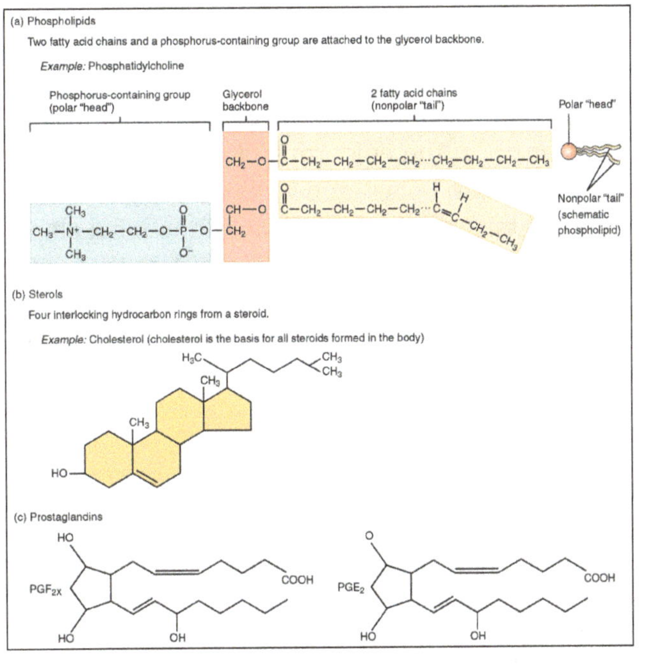

The above figure shows other important lipids: (a) Phospholipids are composed of two fatty acids, glycerol, and a phosphate group. (b) Sterols are ring-shaped lipids. Shown here is cholesterol. (c) Prostaglandins are derived from unsaturated fatty acids. Prostaglandin E2 (PGE2) includes hydroxyl and carboxyl groups.

Steroids

A steroid compound (referred to as a sterol) has as its foundation a set of four hydrocarbon rings bonded to a variety of other atoms and molecules. Although both plants and animals synthesize sterols, the type that makes the most important contribution to human structure and function is cholesterol, which is synthesized by the liver in humans and animals and is also present in most animal-based foods. Like other lipids, cholesterol's hydrocarbons make it hydrophobic; however, it has a polar hydroxyl head that is hydrophilic. Cholesterol is an important component of bile acids, compounds that help emulsify dietary fats. In fact, the word root chole- refers to bile. Cholesterol is also a building block of many hormones, signaling molecules that the body releases to regulate processes at distant sites. Finally, like phospholipids, cholesterol molecules are found in the cell membrane, where their hydrophobic and hydrophilic regions help regulate the flow of substances into and out of the cell.

Prostaglandins

Like a hormone, a prostaglandin is one of a group of signaling molecules, but prostaglandins are derived from unsaturated fatty acids. One reason that the omega-3 fatty acids found in fish are beneficial is that they stimulate the production of certain prostaglandins that help regulate aspects of blood pressure and inflammation, and thereby reduce the risk for heart disease. Prostaglandins also sensitize nerves to pain. One class of pain-relieving medications called nonsteroidal anti-inflammatory drugs (NSAIDs) works by reducing the effects of prostaglandins.

Proteins

You might associate proteins with muscle tissue, but in fact, proteins are critical components of all tissues and organs. A protein is an organic molecule composed of amino acids linked by peptide bonds. Proteins include the keratin in the epidermis of skin that protects underlying tissues, the collagen found in the dermis of skin, in bones, and in the meninges that cover the brain and spinal cord. Proteins are also components of many of the body's functional chemicals, including digestive enzymes in the digestive tract, antibodies, the neurotransmitters that neurons use to communicate with other cells, and the peptide-based hormones that regulate certain body functions (for instance, growth hormone). While carbohydrates and lipids are composed of hydrocarbons and oxygen, all proteins also contain nitrogen (N), and many contain sulfur (S), in addition to carbon, hydrogen, and oxygen.

Microstructure of Proteins

Proteins are polymers made up of nitrogen-containing monomers called amino acids. An amino acid is a molecule composed of an amino group and a carboxyl group, together with a variable side chain. Just 20 different amino acids contribute to nearly all of the thousands of different proteins important in human structure and function. Body proteins contain a unique combination of a few dozen to a few hundred of these 20 amino acid monomers. All 20 of these amino acids share a similar structure. All consist of a central carbon atom to which the following are bonded:

- A hydrogen atom;
- An alkaline (basic) amino group NH_2;
- An acidic carboxyl group COOH;
- A variable group.

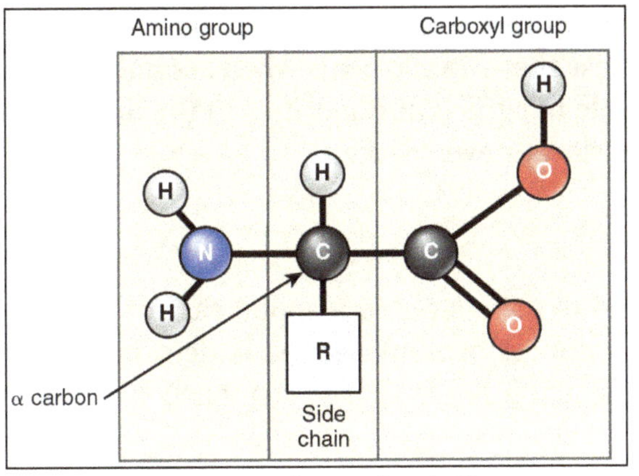

Structure of an Amino Acid

Notice that all amino acids contain both an acid (the carboxyl group) and a base (the amino group) (amine = "nitrogen-containing"). For this reason, they make excellent buffers, helping the body regulate acid–base balance. What distinguishes the 20 amino acids from one another is their variable group, which is referred to as a side chain or an R-group. This group can vary in size and can be polar or nonpolar, giving each amino acid its unique characteristics. For example, the side chains of two amino acids—cysteine and methionine—contain sulfur. Sulfur does not readily participate in hydrogen bonds, whereas all other amino acids do. This variation influences the way that proteins containing cysteine and methionine are assembled.

Amino acids join via dehydration synthesis to form protein polymers. The unique bond holding amino acids together is called a peptide bond. A peptide bond is a covalent bond between two amino acids that forms by dehydration synthesis. A peptide, in fact, is a very short chain of amino acids. Strands containing fewer than about 100 amino acids are generally referred to as polypeptides rather than proteins.

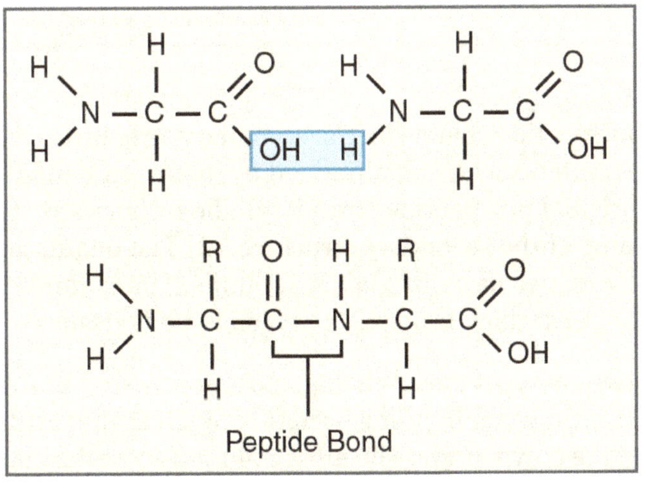

Peptide Bond: Different amino acids join together to form peptides, polypeptides, or proteins via dehydration synthesis. The bonds between the amino acids are peptide bonds.

The body is able to synthesize most of the amino acids from components of other molecules; however, nine cannot be synthesized and have to be consumed in the diet. These are known as the essential amino acids.

Free amino acids available for protein construction are said to reside in the amino acid pool within cells. Structures within cells use these amino acids when assembling proteins. If a particular essential amino acid is not available in sufficient quantities in the amino acid pool, however, synthesis of proteins containing it can slow or even cease.

Shape of Proteins

Just as a fork cannot be used to eat soup and a spoon cannot be used to spear meat, a protein's shape is essential to its function. A protein's shape is determined, most fundamentally, by the sequence of amino acids of which it is made. The sequence is called the primary structure of the protein.

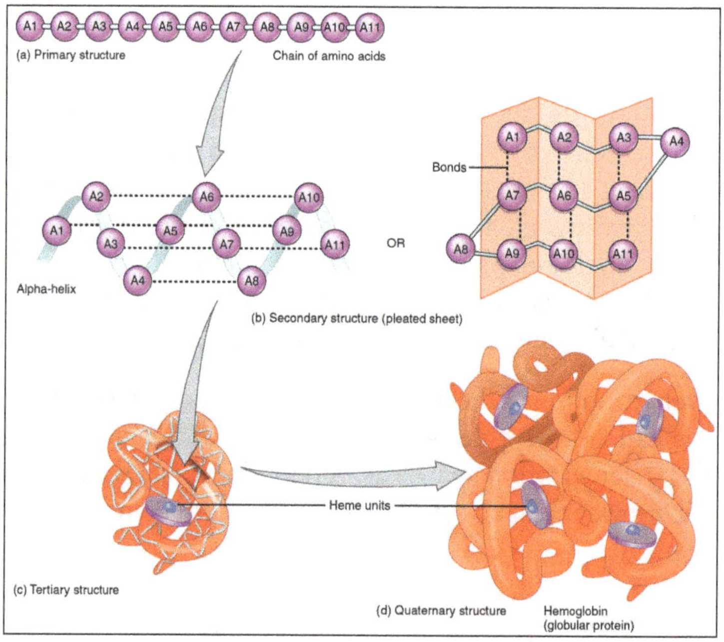

The above figure shows the Shape of Proteins. (a) The primary structure is the sequence of amino acids that make up the polypeptide chain. (b) The secondary structure, which can take the form of an alpha-helix or a beta-pleated sheet, is maintained by hydrogen bonds between amino acids in different regions of the original polypeptide strand. (c) The tertiary structure occurs as a result of further folding and bonding of the secondary structure. (d) The quaternary structure occurs as a result of interactions between two or more tertiary subunits. The example shown here is hemoglobin, a protein in red blood cells which transports oxygen to body tissues.

Although some polypeptides exist as linear chains, most are twisted or folded into more complex secondary structures that form when bonding occurs between amino acids with different properties at different regions of the polypeptide. The most common secondary structure is a spiral called an alpha-helix. If you were to take a length of string and simply twist it into a spiral, it would not hold the shape. Similarly, a strand of amino acids could not maintain a stable spiral shape without the help of hydrogen bonds, which create bridges between different regions of the same strand. Less commonly, a polypeptide chain can form a beta-pleated sheet, in which hydrogen bonds form bridges between different regions of a single polypeptide that has folded back upon itself, or between two or more adjacent polypeptide chains.

The secondary structure of proteins further folds into a compact three-dimensional shape, referred to as the protein's tertiary structure. In this configuration, amino acids that had been very distant in the primary chain can be brought quite close via hydrogen bonds or, in proteins containing cysteine, via disulfide bonds. A disulfide bond is a covalent bond between sulfur atoms in a polypeptide. Often, two or more separate polypeptides bond to form an even larger protein with a quaternary structure. The polypeptide subunits forming a quaternary structure can be identical or different. For instance, hemoglobin, the protein found in red blood cells is composed of four tertiary polypeptides, two of which are called alpha chains and two of which are called beta chains.

When they are exposed to extreme heat, acids, bases, and certain other substances, proteins will denature. Denaturation is a change in the structure of a molecule through physical or chemical means. Denatured proteins lose their functional shape and are no longer able to carry out their jobs. An everyday example of protein denaturation is the curdling of milk when acidic lemon juice is added.

The contribution of the shape of a protein to its function can hardly be exaggerated. For example, the long, slender shape of protein strands that make up muscle tissue is essential to their ability to contract (shorten) and relax (lengthen). As another example, bones contain long threads of a protein called collagen that acts as scaffolding upon which bone minerals are deposited. These elongated proteins, called fibrous proteins, are strong and durable and typically hydrophobic.

In contrast, globular proteins are globes or spheres that tend to be highly reactive and are hydrophilic. The hemoglobin proteins packed into red blood cells are an example; however, globular proteins are abundant throughout the body, playing critical roles in most body functions.

Proteins Function as Enzymes

If you were trying to type a paper, and every time you hit a key on your laptop there was a delay of six or seven minutes before you got a response, you would probably get a new laptop. In a similar way, without enzymes to catalyze chemical reactions, the human body would be nonfunctional. It functions only because enzymes function.

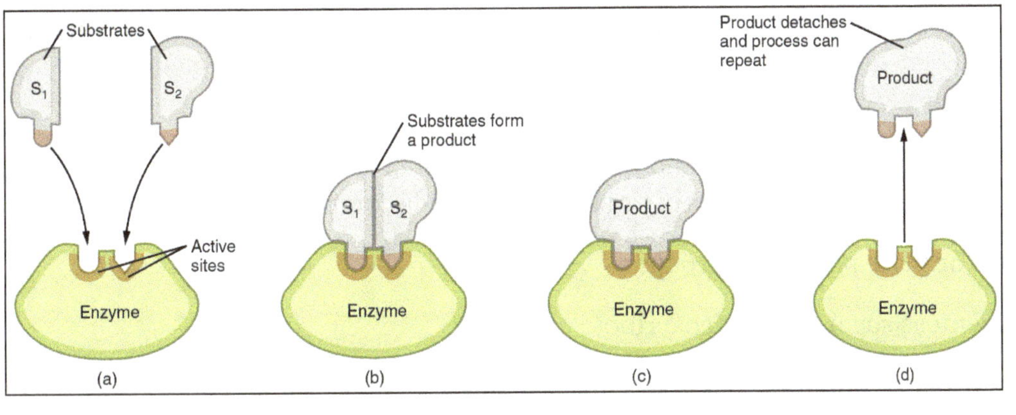

The above image show steps in an Enzymatic Reaction. (a) Substrates approach active sites on enzyme. (b) Substrates bind to active sites, producing an enzyme–substrate complex. (c) Changes internal to the enzyme–substrate complex facilitate interaction of the substrates. (d) Products are released and the enzyme returns to its original form, ready to facilitate another enzymatic reaction.

Enzymatic reactions—chemical reactions catalyzed by enzymes—begin when substrates bind to the enzyme. A substrate is a reactant in an enzymatic reaction. This occurs on regions of the enzyme known as active sites. Any given enzyme catalyzes just one type of chemical reaction. This characteristic, called specificity, is due to the fact that a substrate with a particular shape and electrical charge can bind only to an active site corresponding to that substrate.

Binding of a substrate produces an enzyme–substrate complex. It is likely that enzymes speed up chemical reactions in part because the enzyme–substrate complex undergoes a set of temporary and reversible changes that cause the substrates to be oriented toward each other in an optimal position to facilitate their interaction. This promotes increased reaction speed. The enzyme then releases the products, and resumes its original shape. The enzyme is then free to engage in the process again, and will do so as long as substrate remains.

Other Functions of Proteins

Advertisements for protein bars, powders, and shakes all say that protein is important in building, repairing, and maintaining muscle tissue, but the truth is that proteins contribute to all body tissues, from the skin to the brain cells. Also, certain proteins act as hormones, chemical messengers that help regulate body functions, For example, growth hormone is important for skeletal growth, among other roles.

As was noted earlier, the basic and acidic components enable proteins to function as buffers in maintaining acid–base balance, but they also help regulate fluid–electrolyte balance. Proteins attract fluid, and a healthy concentration of proteins in the blood, the cells, and the spaces between cells helps ensure a balance of fluids in these various "compartments." Moreover, proteins in the cell membrane help to transport electrolytes in and out of the cell, keeping these ions in a healthy balance. Like lipids, proteins can bind with carbohydrates. They can thereby produce glycoproteins or proteoglycans, both of which have many functions in the body.

The body can use proteins for energy when carbohydrate and fat intake is inadequate, and stores of glycogen and adipose tissue become depleted. However, since there is no storage site for protein except functional tissues, using protein for energy causes tissue breakdown, and results in body wasting.

Nucleotides

The fourth type of organic compound important to human structure and function are the nucleotides. A nucleotide is one of a class of organic compounds composed of three subunits:

- One or more phosphate groups.
- A pentose sugar: either deoxyribose or ribose.
- A nitrogen-containing base: adenine, cytosine, guanine, thymine, or uracil.

Nucleotides can be assembled into nucleic acids (DNA or RNA) or the energy compound adenosine triphosphate.

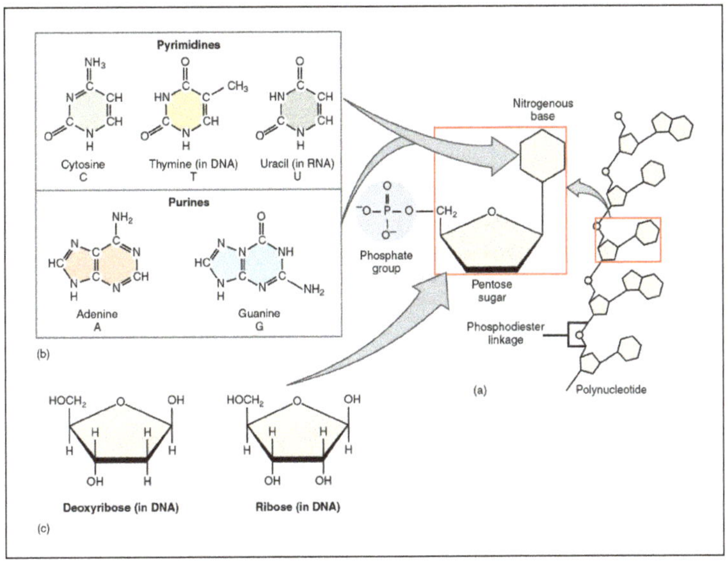

Nucleotides. (a) The building blocks of all nucleotides are one or more phosphate groups, a pentose sugar, and a nitrogen-containing base. (b) The nitrogen-containing bases of nucleotides. (c) The two pentose sugars of DNA and RNA.

Nucleic Acids

The nucleic acids differ in their type of pentose sugar. Deoxyribonucleic acid (DNA) is nucleotide that stores genetic information. DNA contains deoxyribose (so-called because it has one less atom of oxygen than ribose) plus one phosphate group and one nitrogen-containing base. The "choices" of base for DNA are adenine, cytosine, guanine, and thymine. Ribonucleic acid (RNA) is a ribose-containing nucleotide that helps manifest the genetic code as protein. RNA contains ribose, one phosphate group, and one nitrogen-containing base, but the "choices" of base for RNA are adenine, cytosine, guanine, and uracil.

The nitrogen-containing bases adenine and guanine are classified as purines. A purine is a nitrogen-containing molecule with a double ring structure, which accommodates several nitrogen atoms. The bases cytosine, thymine (found in DNA only) and uracil (found in RNA only) are pyramidines. A pyramidine is a nitrogen-containing base with a single ring structure.

Bonds formed by dehydration synthesis between the pentose sugar of one nucleic acid monomer and the phosphate group of another form a "backbone," from which the components' nitrogen-containing bases protrude. In DNA, two such backbones attach at their protruding bases via hydrogen bonds. These twist to form a shape known as a double helix. The sequence of nitrogen-containing bases within a strand of DNA form the genes that act as a molecular code instructing cells in the assembly of amino acids into proteins. Humans have almost 22,000 genes in their DNA, locked up in the 46 chromosomes inside the nucleus of each cell (except red blood cells which lose their nuclei during development). These genes carry the genetic code to build one's body, and are unique for each individual except identical twins.

In contrast, RNA consists of a single strand of sugar-phosphate backbone studded with bases. Messenger RNA (mRNA) is created during protein synthesis to carry the genetic instructions from the DNA to the cell's protein manufacturing plants in the cytoplasm, the ribosomes.

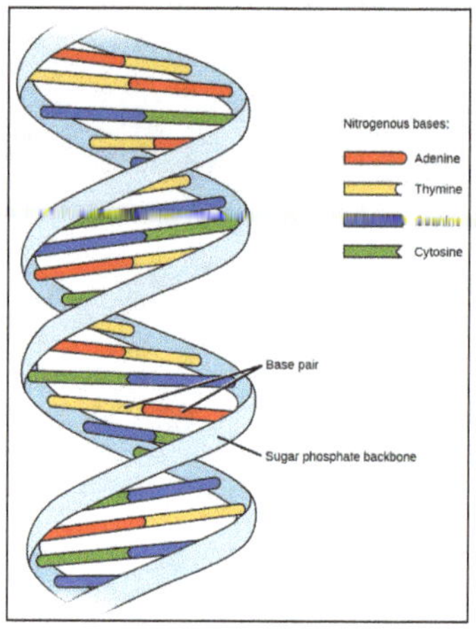

DNA: In the DNA double helix, two strands attach via hydrogen bonds between the bases of the component nucleotides.

Adenosine Triphosphate

The nucleotide adenosine triphosphate (ATP), is composed of a ribose sugar, an adenine base, and three phosphate groups. ATP is classified as a high energy compound because the two covalent bonds linking its three phosphates store a significant amount of potential energy. In the body, the energy released from these high energy bonds helps fuel the body's activities, from muscle contraction to the transport of substances in and out of cells to anabolic chemical reactions.

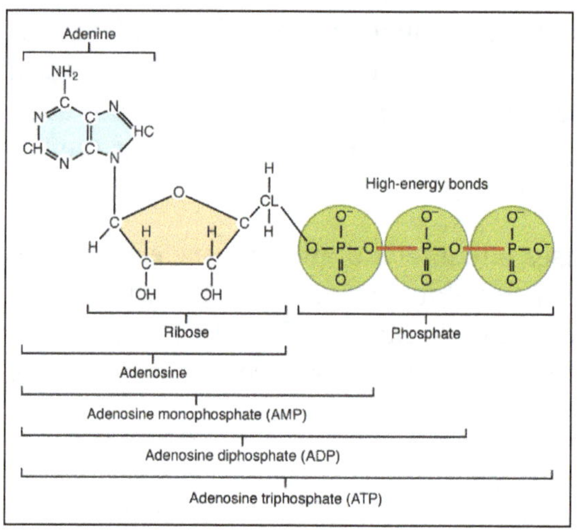

Structure of Adenosine Triphosphate (ATP).

When a phosphate group is cleaved from ATP, the products are adenosine diphosphate (ADP) and inorganic phosphate (P_i). This hydrolysis reaction can be written:

$$ATP + H_2O \rightarrow ADP + P_i + energy$$

Removal of a second phosphate leaves adenosine monophosphate (AMP) and two phosphate groups. Again, these reactions also liberate the energy that had been stored in the phosphate-phosphate bonds. They are reversible, too, as when ADP undergoes phosphorylation. Phosphorylation is the addition of a phosphate group to an organic compound, in this case, resulting in ATP. In such cases, the same level of energy that had been released during hydrolysis must be reinvested to power dehydration synthesis.

Cells can also transfer a phosphate group from ATP to another organic compound. For example, when glucose first enters a cell, a phosphate group is transferred from ATP, forming glucose phosphate ($C_6H_{12}O_6$—P) and ADP. Once glucose is phosphorylated in this way, it can be stored as glycogen or metabolized for immediate energy.

References

- General-introduction-organic-compounds, organic-chemistry, chemistry: toppr.com, Retrieved 12 June, 2019
- Hydrocarbon, science: britannica.com, Retrieved 8 March, 2019
- Saturated-hydrocarbons, chemistry: byjus.com, Retrieved 11 January, 2019
- Hydrocarbon, science: britannica.com, Retrieved 18 July, 2019
- Aromatic-hydrocarbons, chemistry: byjus.com, Retrieved 8 January, 2019
- Organometallic-compounds, chemistry: byjus.com, Retrieved 19 April, 2019
- Metal-clusters, organometallic-compound, science: britannica.com, Retrieved 20 February, 2019
- Heterocyclic-compound, science: britannica.com, Retrieved 21 July, 2019
- Formaldehyde: biologydictionary.net, Retrieved 25 April, 2019
- Weast, Robert C., ed. (1981). CRC Handbook of Chemistry and Physics (62nd ed.). Boca Raton, FL: CRC Press. pp. C–301, E–61. ISBN 0-8493-0462-8
- Gumprecht, W. H. (1968) "3-Bromopyrene" Org. Synth., vol. 48, p. 30. doi:10.15227/orgsyn.048.0030
- Naphthalene: newworldencyclopedia.org, Retrieved 3 May, 2019
- 2-5-organic-compounds-essential-to-human-functioning: opentextbc.ca, Retrieved 17 August, 2019

3

Classification of Organic Reactions

Reactions in organic chemistry are classified into a few broad categories. A few of these are acid-base reactions, addition reactions, substitution reactions and rearrangement reactions. The topics elaborated in this chapter will help in gaining a better perspective about these types of organic reactions.

Organic reactions are the chemical reactions that are undergone by organic compounds.

Organic Reaction Mechanism

Organic Reaction Mechanisms can be identified as a keystone series in organic and synthetic chemistry. Organic reactions are chemical reactions involving organic compounds resulting in a novel compound. Number of possible organic reactions and mechanisms has no limit. However, certain common patterns are discerned that can be used to illustrate many common or helpful reactions. Each reaction has a stepwise reaction mechanism which explains how it happens, even though this comprehensive report of steps is not clear all the time from a list of reactants alone. Organic reactions can be prearranged into quite a lot of basic types. Some reactions fit into more than one group. For instance, a few substitution reactions follow an addition-elimination pathway. This indication isn't proposed to comprise every single organic reaction. Rather, it is premeditated to cover the basic reactions. Elucidating the proposed steps by which the reagents in a reaction mixture are transformed to the observed products takes watchful and conservative interpretation of painstaking experiments.

The mechanism is a presumption, and, like all theories in science, is not static: it becomes more meticulous, or changes entirely, as new experiments are applied to the problem. This theory of how a reaction occurs makes possible an understanding of why it does and thus: prediction of conditions which will assist or hinder the reaction, manipulation of stereochemistry, and prediction of totally new reactions. In addition, the understanding of how and why a reaction occurs provides a framework to aid recall of the outcome of specific reactions.

Acid-base Reactions

An acid-base reaction is a chemical reaction that occurs between an acid and a base. Several concepts exist which provide alternative definitions for the reaction mechanisms involved and their application in solving related problems. Despite several similarities in definitions, their importance becomes apparent as different methods of analysis when applied to acid-base reactions for gaseous or liquid species, or when acid or base character may be somewhat less apparent.

Common Acid-base Theories

Lavoisier Definition

Historically, the first of the scientific concepts of acids and bases was provided by the French chemist Antoine Lavoisier, circa 1776. Lavoisier's knowledge of strong acids was mainly restricted to oxyacids, which tend to contain central atoms in high oxidation states surrounded by oxygen, such as HNO_3 and H_2SO_4, and he was not aware of the true composition of the hydrohalic acids, HCl, HBr, and HI. From his limited knowledge, he defined acids in terms of their content of oxygen, and he named oxygen from Greek words meaning "acid-former".

The Lavoisier definition was held as absolute truth for over 30 years, until the 1810 article and subsequent lectures by Sir Humphry Davy, in which Davy proved the lack of oxygen in hydrogen sulfide (H_2S), hydrogen telluride (H_2Te), and the hydrohalic acids.

Liebig Definition

Around 1838, Justus von Liebig proposed a new definition of acids, based on his extensive works on the chemical composition of organic acids. This finished the doctrinal shift from oxygen-based acids to hydrogen-based acids, started by Davy. According to Liebig, an acid is a hydrogen-containing substance in which the hydrogen could be replaced by a metal. Liebig's definition, while completely empirical, remained in use for almost 50 years until the adoption of the Arrhenius definition.

Arrhenius Definition

Devised after the metal-replacement theory of Justus von Liebig and despite some modifications by later theories, the Arrhenius (pictured) concept remains a simple scientific definition of acid-base reaction character.

The Arrhenius definition of acid-base reactions is a more simplified acid-base concept devised by Svante Arrhenius, which was used to provide a modern definition of bases that followed from his work with Friedrich Wilhelm Ostwald in establishing the presence of ions in aqueous solution in 1884, and led to Arrhenius receiving the Nobel Prize in Chemistry in 1903 for "recognition of the extraordinary services rendered to the advancement of chemistry by his electrolytic theory of dissociation".

As defined at the time of discovery, acid-base reactions are characterized by Arrhenius acids, which dissociate in aqueous solution form hydrogen or the later-termed oxonium (H_3O^+) ions, and Arrhenius bases which form hydroxide (OH^-) ions. More recent IUPAC recommendations now suggest the newer term "hydronium" be used in favor of the older accepted term "oxonium" to illustrate reaction mechanisms such as those defined in the Brønsted-Lowry and solvent system definitions more clearly, with the Arrhenius definition serving as a simple general outline of acid-base character More succinctly, the Arrhenius definition can be surmised as, "Arrhenius acids form hydrogen ions in aqueous solution with Arrhenius bases forming hydroxide ions."

The universal aqueous acid-base definition of the Arrhenius concept is described as the formation of water from hydrogen and hydroxide ions, or hydronium ions and hydroxide ions produced from the dissociation of an acid and base in aqueous solution ($2H_2O \rightarrow OH^- + H_3O^+$), which leads to the definition that in Arrhenius acid-base reactions, a salt and water is formed from the reaction between an acid and a base— in more simple scientific definitions, this form of reaction is called a Neutralization reaction.

$$\text{acid}^+ + \text{base}^- \rightarrow \text{salt} + \text{water}$$

The positive ion from a base can form a salt with the negative ion from an acid. For example, two moles of the base sodium hydroxide (NaOH) can combine with one mole of sulfuric acid (H_2SO_4) to form two moles of water and one mole of sodium sulfate.

$$2NaOH + H_2SO_4 \rightarrow 2H_2O + Na_2SO_4$$

Brønsted-Lowry Definition

The Brønsted-Lowry definition, formulated independently by its two proponents Johannes Nicolaus Brønsted and Martin Lowry in 1923 is based upon the idea of protonation of bases through the de-protonation of acids—more commonly referred to as the ability of acids to "donate" hydrogen ions (H^+) or protons to bases, which "accept" them. In contrast to the Arrhenius definition, the Brønsted-Lowry definition refers to the products of an acid-base reaction as conjugate acids and bases to refer to the relation of one proton, and to indicate that there has been a reaction between the two quantities, rather than a "formation" of salt and water, as explained in the Arrhenius definition.

It defines that in reactions, there is the donation and reception of a proton, which essentially refers to the removal of a hydrogen ion bonded within a compound and its reaction with another compound, and not the removal of a proton from the nucleus of an atom, which would require inordinate amounts of energy not attainable through the simple dissociation of acids. In differentiation from the Arrhenius definition, the Brønsted-Lowry definition postulates that for each acid, there is a conjugate acid and base or "conjugate acid-base pair" that is formed through a complete reaction, which also includes water, which is amphoteric:

1 Bonded hydrogen ions dissociate from the water molecules ($2H_2O$)

2 Hydroxide ion (OH^-) forms the conjugate base

3 Oxonium ion forms a conjugate acid by accepting H^+ ion

$$AH + B \rightarrow BH^+ + A^-$$

General formula for representing Brønsted-Lowry reactions:

$$HCl\ (aq) + H_2O \rightarrow H_3O^+\ (aq) + Cl^-\ (aq)$$

Hydrochloric acid completely reacts with water to form the hydronium and chloride ions:

$$CH_3COOH + NH_3 \rightarrow NH_4^+ + CH_3COO^-$$

Acetic acid reacts incompletely with ammonia, no hydronium ions being produced.

Lewis Definition

The Lewis definition of acid-base reactions, devised by Gilbert N. Lewis in 1923 is an encompassing theory to the Brønsted-Lowry and solvent-system definitions with regards to the premise of a donation mechanism, which conversely attributes the donation of electron pairs from bases and the acceptance by acids, rather than protons or other bonded substances and spans both aqueous and non-aqueous reactions.

$$Ag^+ + 2:NH_3 \rightarrow [H_3N:Ag:NH_3]^+$$

A silver cation reacts as an acid with ammonia which acts as an electron-pair donor, forming an ammonia-silver adduct.

In reactions between Lewis acids and bases, there is the formation of an adduct when the highest occupied molecular orbital (HOMO) of a molecule, such as NH_3 with available lone electron pairs donates lone pairs of electrons to the electron-deficient molecule's lowest unoccupied molecular orbital (LUMO) through a co-ordinate covalent bond; in such a reaction, the HOMO-interacting molecule acts as a base, and the LUMO-interacting molecule acts as an acid. In highly-polar molecules, such as boron trifluoride (BF_3), the most electronegative element pulls electrons towards its own orbitals, providing a more positive charge on the less-electronegative element and a difference in its electronic structure due to the axial or equatorial orbiting positions of its electrons, causing repulsive effects from lone pair-bonding pair (Lp-Bp) interactions between bonded atoms in excess of those already provided by bonding pair-bonding pair (Bp-Bp) interactions. Adducts involving metal ions are referred to as co-ordination compounds.

Solvent-system Definition

This definition is based on a generalization of the earlier Arrhenius definition to all autodissociating solvents. In all such solvents there is a certain concentration of a positive species, solvonium cations and negative species, solvate anions, in equilibrium with the neutral solvent molecules. For example:

$$2\,H_2O \rightleftharpoons H_3O^+ (\text{hydronium}) + OH^- (\text{hydroxide})$$

$$2\,NH_3 \rightleftharpoons NH_4^+ (\text{ammonium}) + NH_2^- (\text{amide})$$

In some aprotic systems,

$$N_2O_4 \rightleftharpoons NO^+ (\text{nitrosonium}) + NO_3^- (\text{nitrate})$$
$$2\,SbCl_3 \rightleftharpoons SbCl_2^+ (\text{dichloroantimonium}) + SbCl_4^- (\text{tetrachloroantimonate})$$

A solute causing an increase in the concentration of the solvonium ions and a decrease in the solvate ions is an acid and one causing the reverse is a base. Thus, in liquid ammonia, KNH_2 (supplying NH_2^-) is a strong base, and NH_4NO_3 (supplying NH_4^+) is a strong acid. In liquid sulfur dioxide (SO_2), thionyl compounds (supplying SO^{2+}) behave as acids, and sulfites (supplying SO_3^{2-}) behave as bases.

Here are some nonaqueous acid-base reactions in liquid ammonia:

$$2\,NaNH_2 (\text{base}) + Zn(NH_2)_2 (\text{amphiphilic amide}) \rightarrow Na_2[Zn(NH_2)_4]$$
$$2\,NH_4I (\text{acid}) + Zn(NH_2)_2 (\text{amphiphilic amide}) \rightarrow [Zn(NH_3)_4]I_2$$

Nitric acid can be a base in liquid sulfuric acid:

$$HNO_3 (\text{base}) + 2\,H_2SO_4 \rightarrow NO_2^+ + H_3O^+ + 2\,HSO_4^-$$

And things become even stranger in the aprotic world, for example in liquid N_2O_4:

$$AgNO_3 (\text{base}) + NOCl (\text{acid}) \rightarrow N_2O_4 + AgCl$$

Since solvent-system definition depends on the solvent as well as on the compound itself, the same compound can change its role depending on the choice of the solvent. Thus, $HClO_4$ is a strong acid in water, a weak acid in acetic acid, and a weak base in fluorosulfonic acid.

Usanovich Definition

The most general definition is that of the Russian chemist Mikhail Usanovich, and can basically be summarized as defining an acid as "anything that accepts negative species or donates positive ones," and a base as the reverse. This tends to overlap the concept of redox (oxidation-reduction), and so is not highly favored by chemists. This is because redox reactions focus more on physical electron transfer processes, rather than bond making/bond breaking processes, although the distinction between these two processes is somewhat ambiguous.

Lux-Flood Definition

This definition, proposed by German chemist Hermann Lux in 1939, further improved by Håkon Flood circa 1947 and now commonly used in modern geochemistry and electrochemistry of molten salts, describes an acid as an oxide ion acceptor and a base as an oxide ion donor. For example:

$$MgO (\text{base}) + CO_2 (\text{acid}) \rightarrow MgCO_3$$
$$CaO (\text{base}) + SiO_2 (\text{acid}) \rightarrow CaSiO_3$$
$$NO_3^- (\text{base}) + S_2O_7^{2-} (\text{acid}) \rightarrow NO_2^+ + 2\,SO_4^{2-}$$

Addition Reactions

An addition reaction is a chemical reaction wherein two or more reactants come together to form a larger single product. But only chemical compounds containing multiple bond character can undergo an addition reaction as a double or triple bond is usually broken to form the required single bonds. An addition reaction is essentially a reverse decomposition reaction wherein a decomposition reaction is a reaction where one compounds one or more elements or compounds. Looking at an example of an addition reaction, hydrochlorination of propane (an alkene), for which the equation is:

$$CH_3CH=CH_2 + HCl \rightarrow CH_3C^+HCH_3 + Cl^- \rightarrow CH_3CHClCH_3$$

Rules for Addition Reactions

The following three rules can be used when working with addition reactions:

1. Addition reactions are associated with unsaturated compounds. An unsaturated compound is a compound that contains double or triple bonds. Always, always, remember that you will most likely encounter an addition reaction when your starting compound, or reactant, is unsaturated.
2. When completed, addition reactions lead to a single compound, which is the final product. What you won't find are any leftovers or doggy bags. In chemistry, leftovers are called residues. Addition reactions will not have any reactant residues on the product side.
3. Atoms can only be added to molecules when forming a product; nothing is eliminated or taken away. Remember that addition reactions involve the addition of an atom to a compound containing a single or double bond. Whatever the scenario, adding atoms or combining reactants, addition reactions will never involve the elimination or removal of atoms or molecules.

Types of Addition Reactions

For polar addition reactions there are two classifications, namely:

1. Electrophilic Addition reactions
2. Nucleophilic Addition reactions

For non-polar addition reactions, we have two classifications, namely:

1. Free radical addition reactions
2. Cycloadditions reactions

Electrophilic Addition

An electrophilic addition reaction can be described as an addition reaction in which a reactant with multiple bonds as in a double or triple bond undergoes has its π bond broken and two new σ bonds are formed.

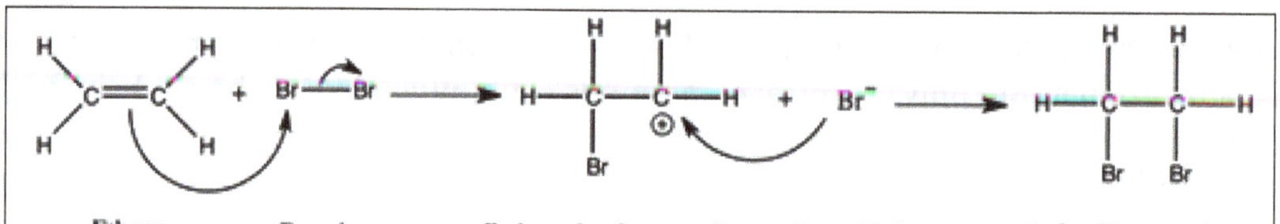

Electrophilic Addition mechanism between Ethene and Bromine.

Nucleophilic Addition

A nucleophilic addition reaction is an addition reaction where a chemical compound with an electron-deficient or electrophilic double or triple bond, a π bond, reacts with a nucleophile which is an electron-rich reactant with the disappearance of the double bond and creation of two new single, or σ, bonds.

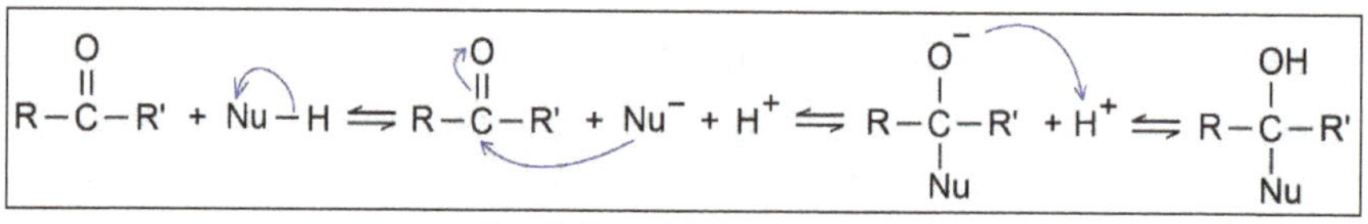

Nucleophilic addition – Ketone.

Substitution Reaction

The substitution reaction is defined as a reaction in which the functional group of one chemical compound is substituted by another group or it is a reaction which involves the replacement of one atom or a molecule of a compound with another atom or molecule.

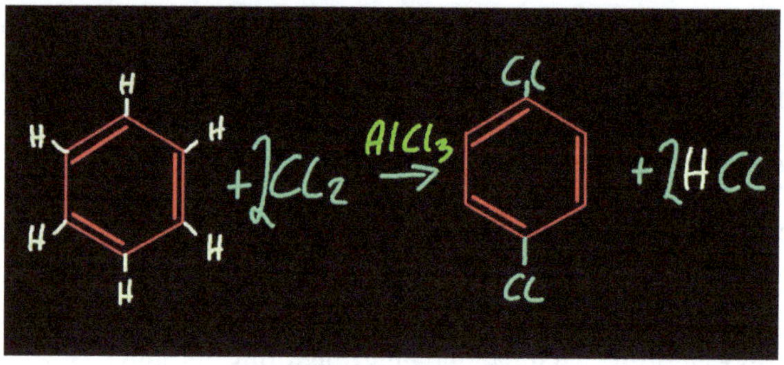

Substitution Reaction

These type of reactions are said to possess primary importance in the field of organic chemistry. For example, when CH_3Cl is reacted with the hydroxyl ion (OH^-), it will lead to the formation of the original molecule called methanol with that hydroxyl ion. The following reaction is as shown below:

$$CH_3Cl + (\text{-}OH) \rightarrow CH_3OH(\text{ methanol}) + Cl^-$$

One more example would be the reaction of Ethanol with the hydrogen iodide which forms iodoethane along with water. The reaction is as shown:

$$CH_3CH_2OH + HI \rightarrow CH_3CH_2OI + H_2O$$

In order for substitution reaction to occur there are certain conditions that have to be used. They are:

- Maintaining low temperatures such as room temperature.
- The strong base such as NaOH has to be in dilute form. Suppose if the base is of a higher concentration, there are chances of dehydrohalogenation taking place.
- The solution needs to be in an aqueous state such as water.

Types of Substitution Reactions

Substitution Reactions are of two types naming nucleophilic reaction and electrophilic reactions. These two types of reactions mainly differ in the kind of atom which is attached to its original molecule. In the nucleophilic reactions the atom is said to be electron-rich species, whereas, in the electrophilic reaction, the atom is an electron-deficient species.

Electrophilic Substitution Reaction

An electrophilic substitution reaction is a chemical reaction in which the functional group attached to a compound is replaced by an electrophile. The displaced functional group is typically a hydrogen atom. Electrophilic substitution reactions generally proceed via a three-step mechanism that involves the following steps:

- The generation of an electrophile.
- The formation of a carbocation (which is an intermediate).
- The removal of a proton from the intermediate.

Types of Electrophilic Substitution Reactions

The two primary types of electrophilic substitution reactions undergone by organic compounds are electrophilic aromatic substitution reactions and electrophilic aliphatic substitution reactions. An illustration describing the electrophilic substitution of a hydrogen atom (belonging to a benzene molecule) with a chlorine atom is provided below:

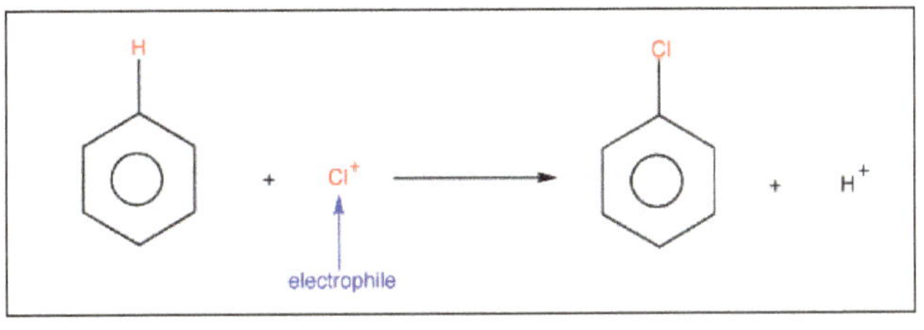

Here, the chlorine cation acts as an electrophile and replaces a hydrogen atom in the benzene ring. The products formed in this electrophilic substitution reaction include a proton and a chlorobenzene molecule.

Electrophilic Aromatic Substitution Reaction

In electrophilic aromatic substitution reactions, an atom attached to an aromatic ring is replaced with an electrophile. Examples of such reactions include aromatic nitrations, aromatic sulfonations, and Friedel-Crafts reactions.

It is important to note that the aromaticity of the aromatic compound is preserved in electrophilic aromatic substitutions. Therefore, these reactions can be used to obtain aryl halides from aromatic rings and iodine, bromine, or chlorine.

Electrophilic Aliphatic Substitution Reaction

In electrophilic aliphatic substitution reactions, an electrophile replaces the functional group (generally hydrogen) in an aliphatic compound. These reactions can be classified into the following five types.

- Halogenation of ketones
- Nitrosation
- Keto-Enol tautomerism
- Insertion of a carbene into a carbon-hydrogen bond
- Diazonium coupling (aliphatic)

These electrophilic substitution reactions can result in an inversion of configuration if the electrophilic attack occurs at an angle of 180° to the leaving group (attack from the rear).

Mechanism of Electrophilic Substitution Reaction

The electrophilic substitution reaction mechanism involves three steps:

Step 1: Generation of Electrophile

Anhydrous aluminium chloride is a very useful Lewis acid in the generation of electrophile from the chlorination, alkylation, and acylation of an aromatic ring. The resulting electrophiles (from the combination of anhydrous aluminium chloride and the attacking reagent) are Cl^+, R^+, and RC^+O respectively as shown below:

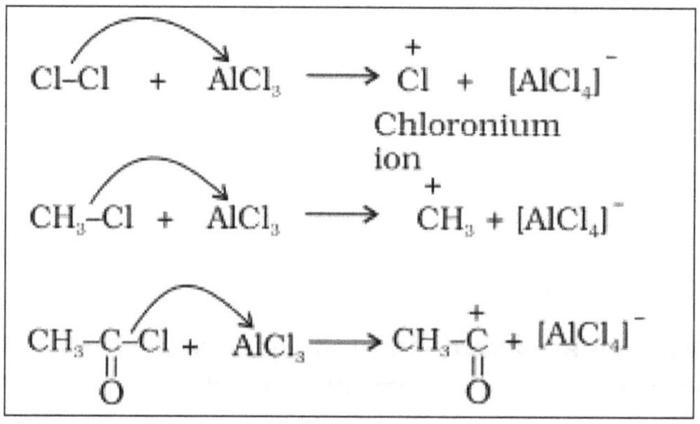

Step 2: Formation of Carbocation

The electrophile attacks the aromatic ring, forming a sigma complex or an arenium ion. One of the carbons in this arenium ion is sp^3 hybridized.

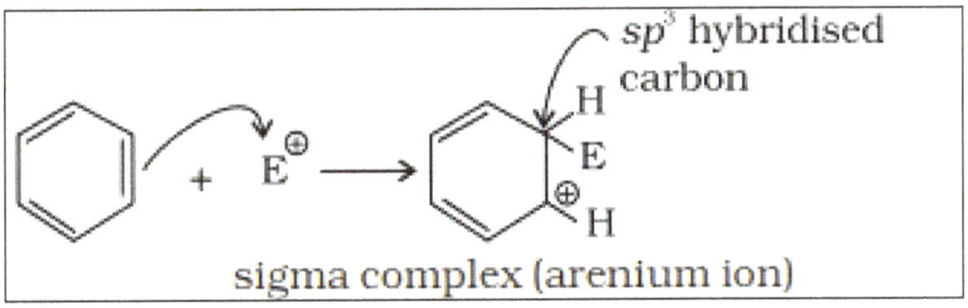

This arenium ion finds stability in a resonance structure. Since the delocalization of electrons stops at the sp^3 hybridized carbon, the sigma complex or the arenium ion loses its aromatic character.

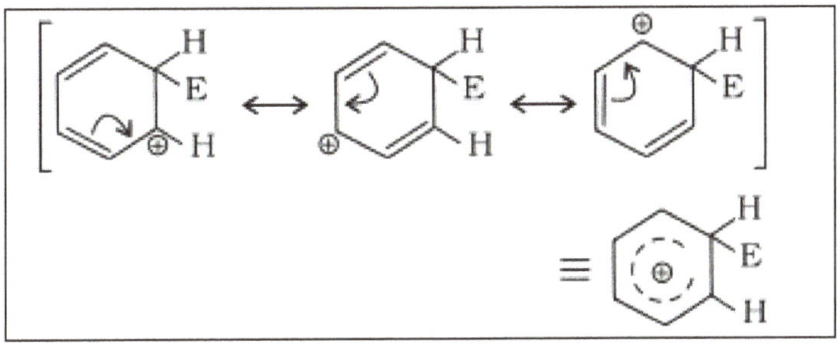

Step 3: Removal of Proton

In order to restore the aromatic character, the sigma complex releases a proton from the sp^3 hybridized carbon when it is attacked by the $[AlCl_4]^-$. The reaction describing the removal of a proton from the sigma complex is given below:

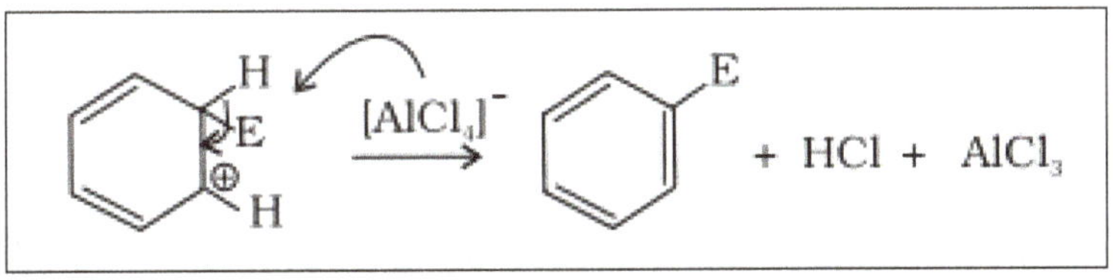

Thus, the electrophile replaces the hydrogen atom in the benzene ring. The electrophilic substitution reaction is a very important reaction in organic chemistry as the concept is used in many organic name reactions.

Nucleophilic Substitution Reaction

Nucleophilic substitution reactions occur when an electron rich species, the nucleophile, reacts at an electrophilic saturated C atom attached to an electronegative group (important), the leaving group, that can be displaced as shown by the general scheme. The electrophilic C can be recognised by looking for the polar σ bond due to the presence of an electronegative substituent (C-Cl, C-Br, C-I and C-O). Nucleophilic substitution reactions are an important class of reactions that allow the interconversion of functional groups. Of particular importance are the reactions of alkyl halides (R-X) and alcohols

(R-OH). For alcohols, the range of substitution reactions possible can be increased by utilising the tosylates (R-OTs), an alternative method of converting the -OH to a better leaving group.

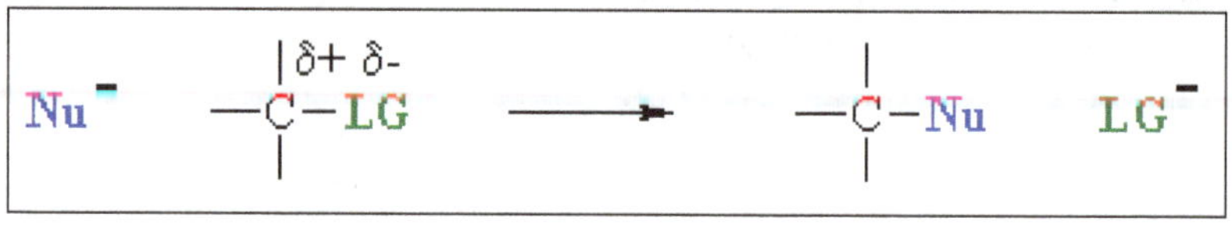

There are two mechanistic models for how a nucleophilic substitution reaction can proceed. In the first picture, the reaction takes place in a single step, and bond-forming and bond-breaking occur simultaneously.

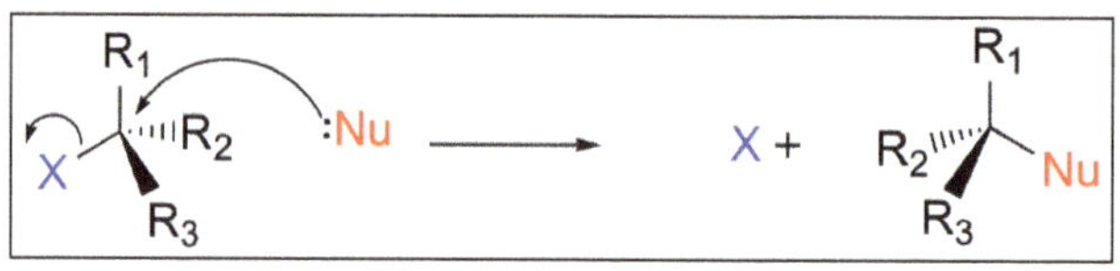

This is called an 'associative', or 'S_N2' mechanism. In the term S_N2, S stands for 'substitution', the subscript N stands for 'nucleophilic', and the number 2 refers to the fact that this is a bimolecular reaction: the overall rate depends on a step in which two separate molecules (the nucleophile and the electrophile) collide. A potential energy diagram for this reaction shows the transition state (TS) as the highest point on the pathway from reactants to products.

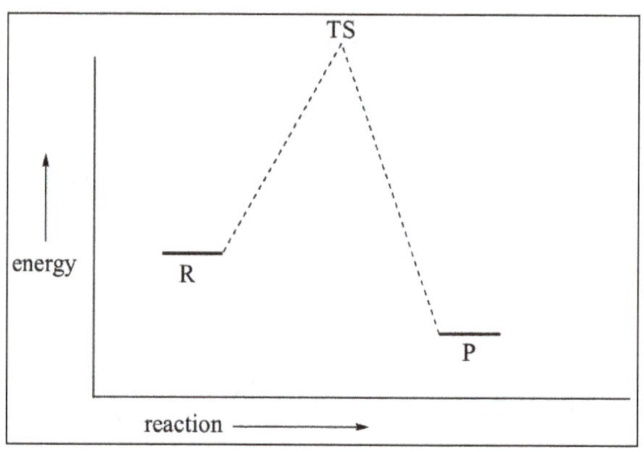

If you look carefully at the progress of the S_N2 reaction, you will realize something very important about the outcome. The nucleophile, being an electron-rich species, must attack the electrophilic carbon from the *back side* relative to the location of the leaving group. Approach from the front side simply doesn't work: the leaving group - which is also an electron-rich group - blocks the way.

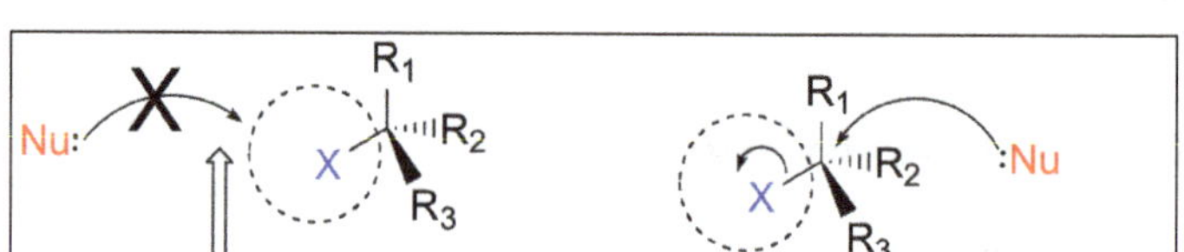

nucleophilic attack is blocked from the front side . . .

. . .so attack occurs from the back side

The result of this backside attack is that the stereochemical configuration at the central carbon *inverts* as the reaction proceeds. In a sense, the molecule is turned inside out. At the transition state, the electrophilic carbon and the three 'R' substituents all lie on the same plane.

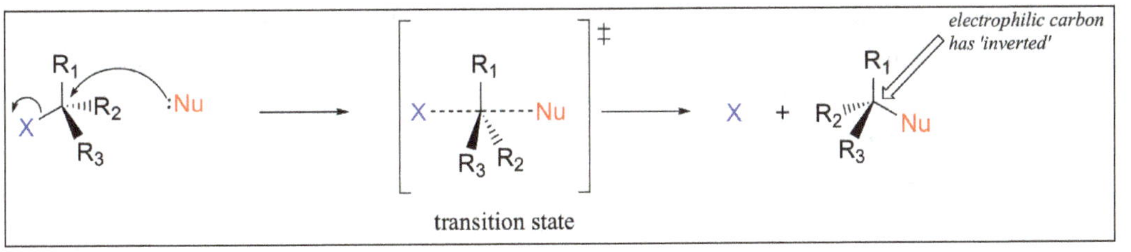

What this means is that S_N2 reactions whether enzyme catalyzed or not, are inherently stereoselective: when the substitution takes place at a stereocenter, we can confidently predict the stereochemical configuration of the product.

Dissociative Nucleophilic Substitution: The S_N1 Reaction

A second model for a nucleophilic substitution reaction is called the 'dissociative', or 'S_N1' mechanism: in this picture, the C-X bond breaks *first*, before the nucleophile approaches:

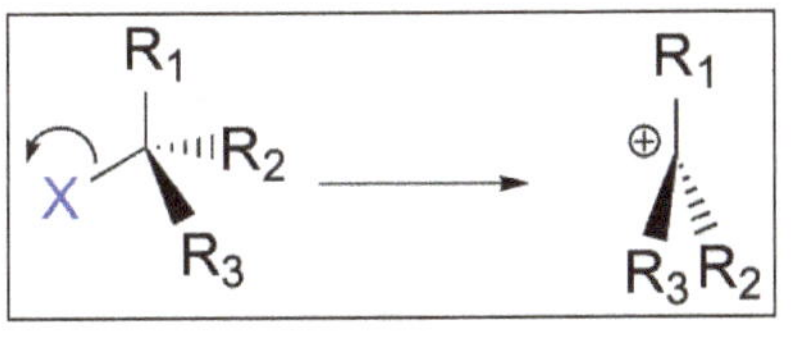

This results in the formation of a carbocation because the central carbon has only three bonds, it bears a formal charge of +1. Recall that a carbocation should be pictured as sp^2 hybridized, with trigonal planar geometry. Perpendicular to the plane formed by the three sp^2 hybrid orbitals is an empty, unhybridized *p* orbital.

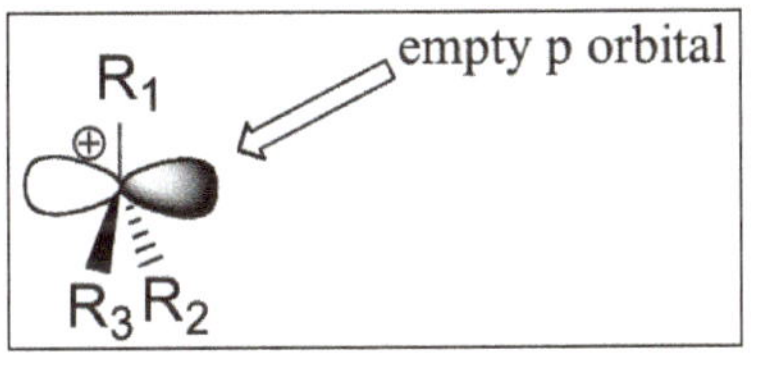

In the second step of this two-step reaction, the nucleophile attacks the empty, 'electron hungry' *p* orbital of the carbocation to form a new bond and return the carbon to tetrahedral geometry.

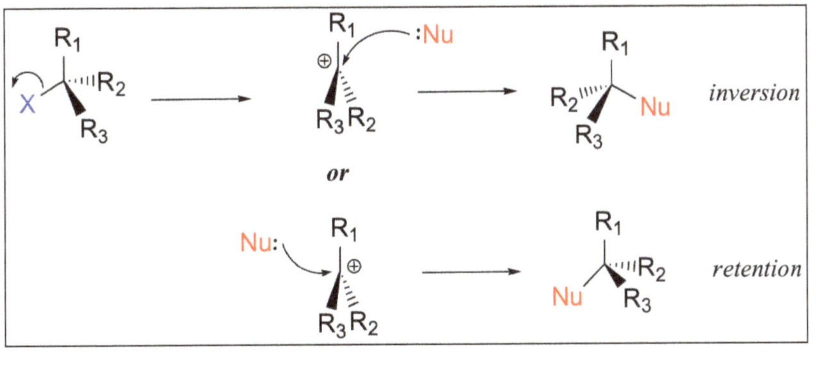

We saw that S_N2 reactions result specifically in inversion of stereochemistry at the electrophilic carbon center. What about the stereochemical outcome of S_N1 reactions? In the model S_N1 reaction shown above, the leaving group dissociates completely from the vicinity of the reaction before the nucleophile begins its attack. Because the leaving group is no longer in the picture, the nucleophile is free to attack from either side of the planar, sp^2-hybriduzed carbocation electrophile. This means that about half the time the product has the same stereochemical configuration as the starting material (retention of configuration), and about half the time the stereochemistry has been inverted. In other words, racemization has occurred at the carbon center. As an example, the tertiary alkyl bromide below would be expected to form a racemic mix of R and S alcohols after an S_N1 reaction with water as the incoming nucleophile.

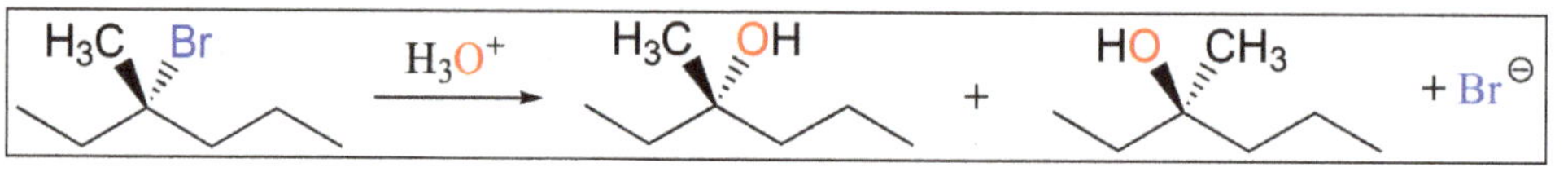

While most nonenzymatic S_N1 reactions are not stereoselective, Enzyme-catalyzed nucleophilic substitution reactions - whether S_N1 or S_N2 - almost always *are* stereoselective. The direct result of an enzymatic nucleophilic substitution reaction is more often than not *inversion* of configuration - this is because the leaving group usually remains bound in the enzyme's active site long enough to block a nucleophilic attack from that side. This does not mean, however, that enzymes can only catalyze substitution reactions with inversion of configuration.

In the S_N1 reaction we see an example of a reaction intermediate, a very important concept in the study of organic reaction mechanisms. Recall that many important organic reactions do not occur in a single step; rather, they are the sum of two or more discreet bond-forming/bond-breaking steps, and involve transient intermediate species that go on to react very quickly. In the S_N1 reaction, the carbocation species is a reaction intermediate. A potential energy diagram for an S_N1 reaction shows that the carbocation intermediate can be visualized as a kind of valley in the path of the reaction, higher in energy than both the reactant and product but lower in energy than the two transition states.

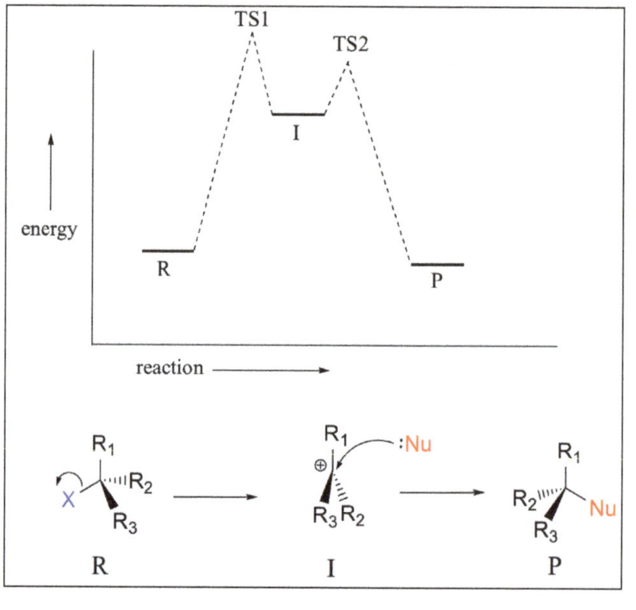

The first step of the reaction above, in which two charged species are formed from a neutral molecule, is much the slower of the two steps, and is therefore rate-determining. This is illustrated by the energy diagram, where the activation energy for the first step is higher than that for the second step. Also recall that an S_N1 reaction has *first order* kinetics, because the rate determining step involves one molecule splitting apart, not two molecules colliding.

Many S_N1 reactions are of a class that are referred to as solvolysis, where a solvent molecule participates in the reaction as a nucleophile. The S_N1 reaction of allyl bromide in methanol is an example of what we would call methanolysis, while if water is the solvent the reaction would be called hydrolysis:

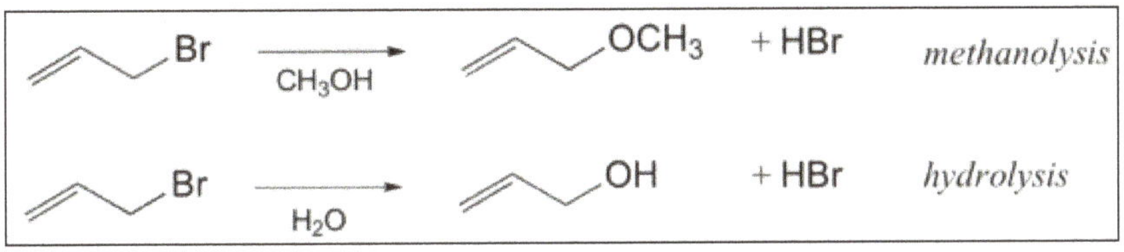

Because water and alcohols are relatively weak nucleophiles, they are unlikely to react in an S_N2 fashion.

Nucleophilic Substitutions Occur at sp³-hybridized Carbons

One more important point must be made before continuing: nucleophilic substitutions as a rule occur at sp^3-hybridized carbons, and *not* where the leaving group is attached to an sp^2-hybridized carbon:

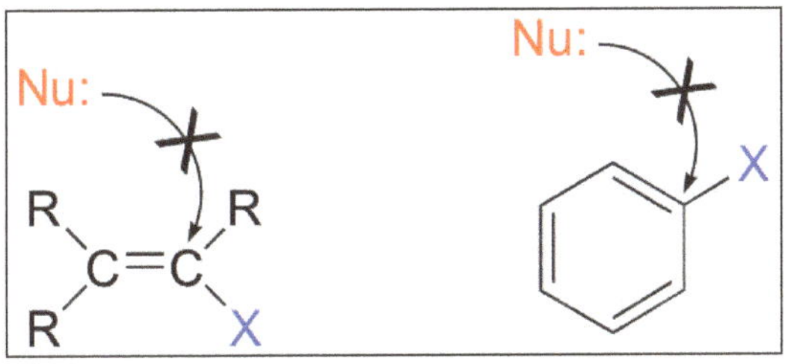

Bonds on sp^2-hybridized carbons are inherently shorter and stronger than bonds on sp^3-hybridized carbons, meaning that it is harder to break the C-X bond in these substrates. S_N2 reactions of this type are unlikely also because the (hypothetical) electrophilic carbon is protected from nucleophilic attack by electron density in the p bond. S_N1 reactions are highly unlikely, because the resulting carbocation intermediate, which would be sp-hybridized, would be very unstable.

Elimination Reactions

Elimination reaction often competes with substitution reactions. In this reaction, a substrate (typically an alkyl halide) eliminates one equivalent (unit) of acid to form an alkene. Two possible mechanisms are available for this elimination reaction – E1 and E2 mechanisms.

Elimination reaction is a type of reaction is mainly used to transform saturated *compounds* (organic compounds which contain single carbon-carbon bonds) to unsaturated compounds (compounds which feature double or triple carbon-carbon bonds). Besides, it is an important method for the preparation of alkenes.

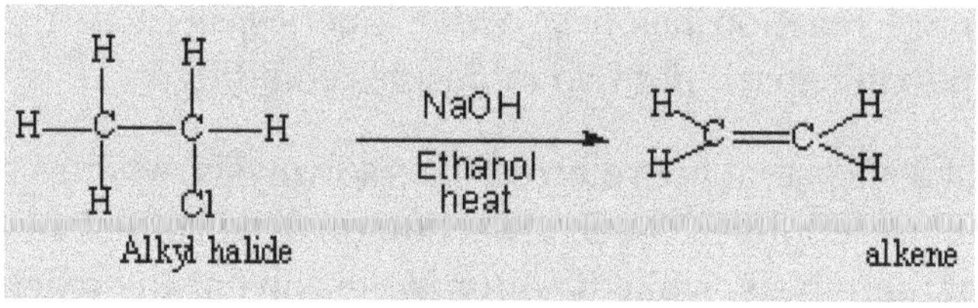

Preparation of Alkenes

An elimination reaction is a type of a chemical reaction where several atoms either in pairs or groups are removed from a molecule. The removal usually takes place due to the action of acids and bases or action of metals. It can also happen through the process of heating at high temperatures.

Important Methods of Elimination Reaction

Normally, elimination reactions are distinguished by the kind of atoms or groups of atoms that leave the molecule. Due to this, there are two main methods involved in this type of reaction;

- Dehydration
- Dehydrohalogenation

In the dehydration method, there is the elimination of a water molecule mostly from compounds such as alcohol. Sometimes, this method is also called Beta elimination reaction where the leaving group and H are placed at neighbour carbon atoms. On the other hand, in dehydrohalogenation, there is a removal of a hydrogen atom and a halogen atom.

Some of the other common types of elimination reactions are α-elimination and γ- and δ-elimination.

Mechanism of Elimination Reaction

The elimination reaction consists of three fundamental events and they are;

- Proton removal.
- C-C pi bond is formed.
- There is a breakage in the bond of the leaving group.

Depending on the reaction kinetics, elimination reactions can occur mostly by two mechanisms namely E1 or E2 where E is referred to as elimination and the number represent the molecularity.

E1 Reaction

- In the E1 mechanism which is also known as unimolecular elimination, there are usually two steps involved – ionization and deprotonation.
- During ionization, there is a formation of carbocation as an intermediate. In deprotonation, a proton is lost by the carbocation.
- This happens in the presence of a base which further leads to the formation of a pi-bond in the molecule.

- In E1, the reaction rate is also proportional to the concentration of the substance to be transformed.
- It exhibits first-order kinetics.

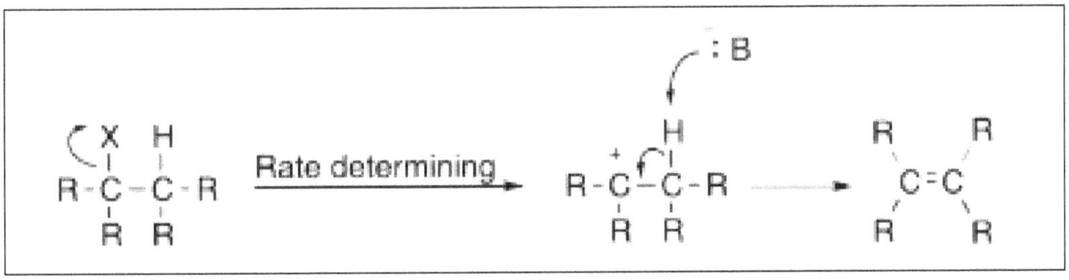

E1 mechanism shares the features of the SN1 reaction. The initial step is the formation of a carbocation intermediate through the loss of the leaving group. This slow step becomes the rate-determining step for the whole reaction.

E2 Reaction

- In an E2 mechanism which refers to *bimolecular elimination* is basically a one-step mechanism.
- Here, the carbon-hydrogen and carbon-halogen bonds mostly break off to form a new double bond.
- However, in the E2 mechanism, a base is part of the rate determining step and it has a huge influence on the mechanism.
- The reaction rate is mostly proportional to the concentrations of both the eliminating agent and the substrate.
- It exhibits second-order kinetics.

The E2 mechanism can generally be represented as below. In the below-mentioned representation, B stands for base and X stands for the halogen.

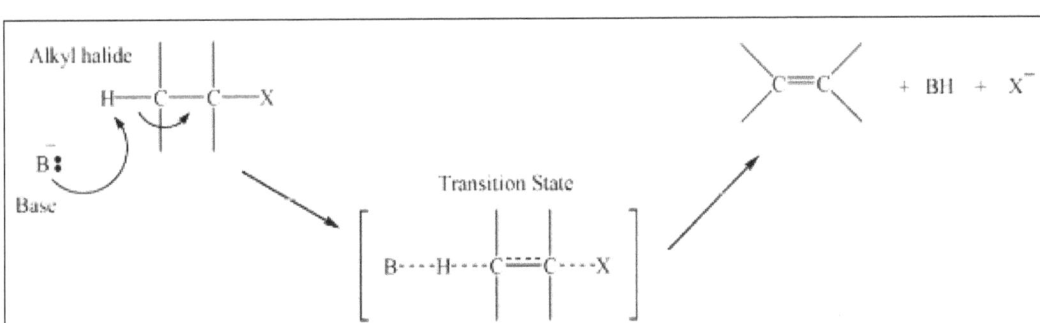

The rate of the E2 reaction is,

$$\text{Rate} = k[\text{RX}][\text{Base}]$$

So the reaction rate depends on both the substrate (RX) and the base involved. In the elimination reaction, the major product formed is the most stable alkene.

One of the common examples of elimination reaction is the dehydration of alcohol. Here the process takes place in the presence of a base such as an ethoxide ion ($C_2H_5O^-$). It can be represented as;

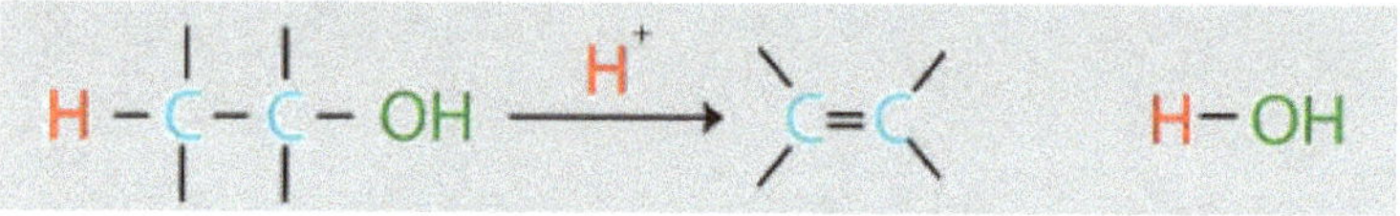

Elimination Reaction Mechanism

"E2 and E1 reactions differ significantly in the nature of the transition states that determine the regiochemistry of the product". The E2 pathway involves a transition state leading from starting material directly to the product. The product forming step of an E1 reaction is more exothermic than that of an E2 reaction. Thus the E1 reaction has a relatively early transition state, closely resembling the carbocation formed in the rate determining step.

Rearrangement Reactions

A reaction in which an atom or bond moves or migrates, having been initially located at one site in a reactant molecule and ultimately located at a different site in a product molecule. A rearrangement reaction may involve several steps, but the key feature defining it as a rearrangement is that a bond shifts from one site of attachment to another. The simplest examples of rearrangement reactions are intramolecular, that is, reactions in which the product is simply a structural isomer of the reactant.

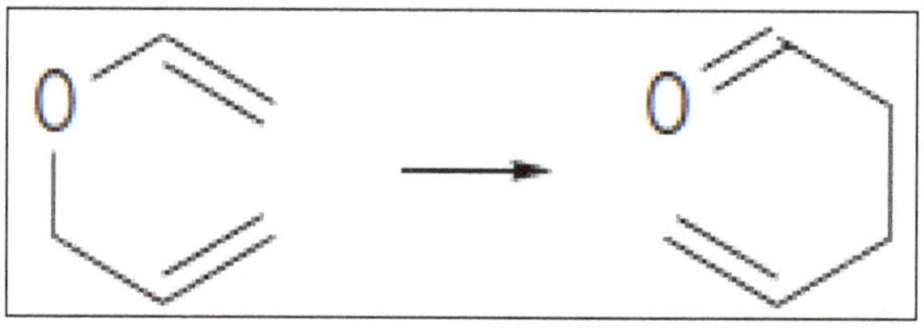

Types of Rearrangements

1. Rearrangement to Electron Deficient Carbon:

- Carbon Migration
  - Wagner-Meerwein Rearrangement
  - Pinacol Rearrangement
  - Benzilic Acid Rearrangement
  - Arndt-Eistert Homologation Reaction.
- Halogen, Oxygen, Sulfur and Nitrogen Migration

2. Rearrangement to Electron Deficient Nitrogen:

- Hofmann Rearrangement
- Curtius Rearrangement
- Schmidt Rearrangement
- Lossen Rearrangement
- Beckmann Rearrangement

3. Rearrangement to Electron Deficient Oxygen:

- Baeyer Villiger Reaction
- Hydroperoxide Rearrangement
- Dakin Reaction

4. Rearrangement to Electron-Rich Carbon:

- Stevens Rearrangement
- Sommelet-Hauser Rearrangement
- Wittig Rearrangement
- Favorskii Rearrangement

5. Aromatic Rearrangements:

- Intermolecular Migration from Nitrogen to Carbon
- Fries Rearrangement
- Intramolecular Migration from Nitrogen to Carbon
- Claisen Rearrangement

Rearrangement to Electron Deficient Carbon

Carbon Migration

Wagner-Meerwein Rearrangement: It is one of the simplest systems where an alkyl group migrates, with its bonding pair, to an electron-dcficient carbon atom.

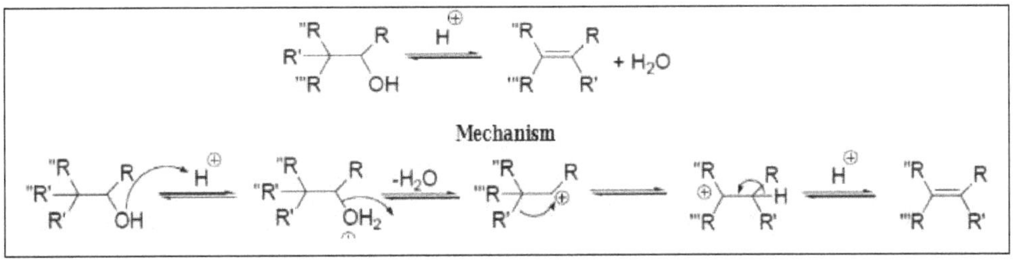

Pinacol Rearrangement: Treatment of 1,2-diols (pinacol) with acid lead to rearrangement to give ketone.

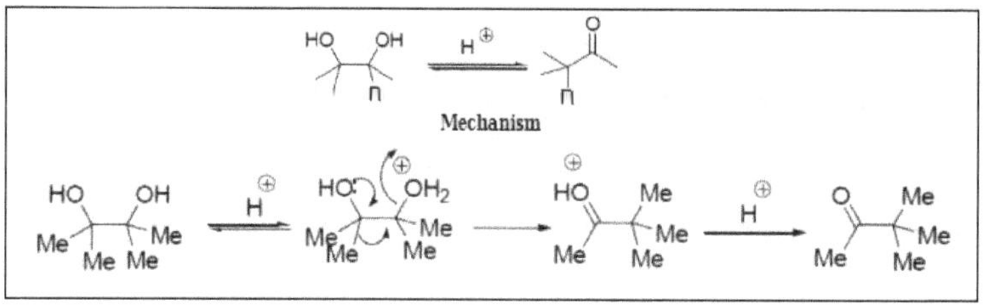

Halogen, Oxygen, Sulfur and Nitrogen Migration

In the system X-C-C-Y, an atom X with an unshared pair of electrons can assist the heterolysis of the C-Y bond. In case of unsymmetrical system, nucleophilic attack predominates at the less substituted carbon of the bridged ion that leads to rearranged skeleton.

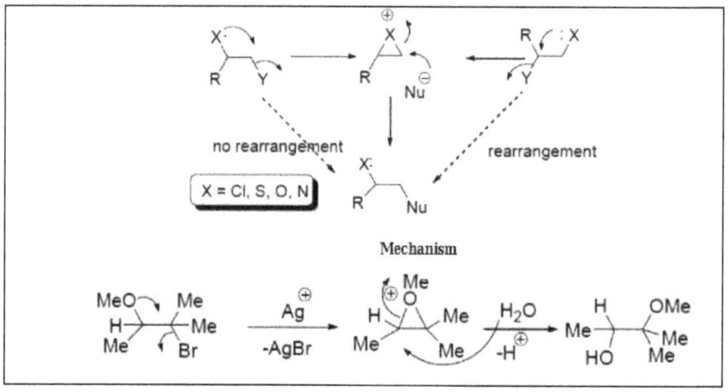

Rearrangement to Electron Deficient Nitrogen

Hofmann Rearrangement - This rearrangement provides an effective method for the synthesis of primary aliphatic and aromatic amines from primary amides.

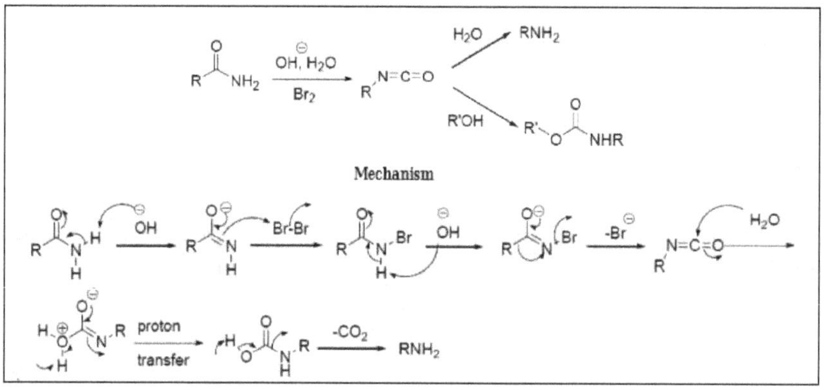

Rearrangement to Electron Deficient Oxygen

Dakin Reaction - Benzaldehyde or acetophenone bearing hydroxyl substituent in the ortho or para position proceed rearrangement to give catechol or quinol, respectively.

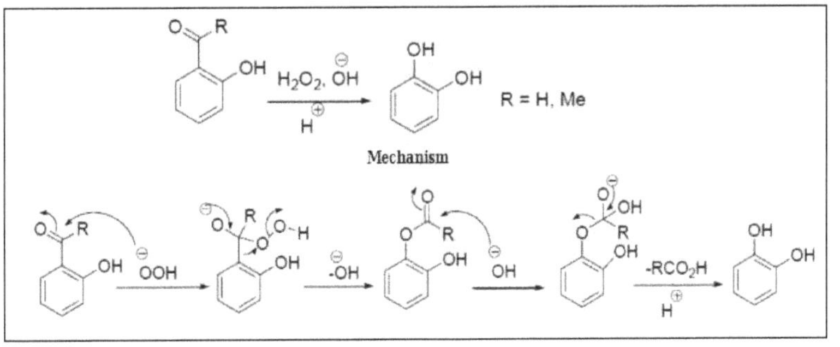

Rearrangement to Electron-rich Carbon

This group of reaction has been less explored, and is less of synthetic importance compared to the rearrangements to electron deficient carbons. The rearrangements to electron deficient hetero atom may be generally explained as:

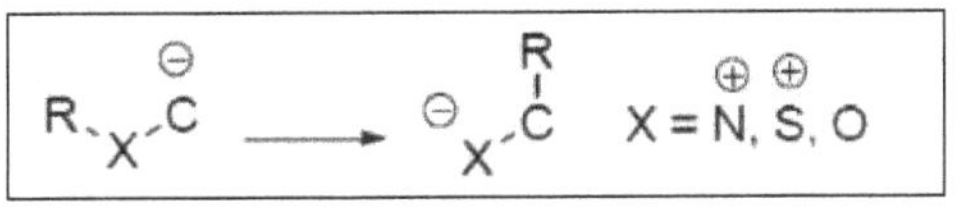

Wittig Rearrangement - Ethers undergo [1,2]-sigmatropic rearrangement in the presence of strong base such as amide ion or phenyllithium to give more stable oxyanion.

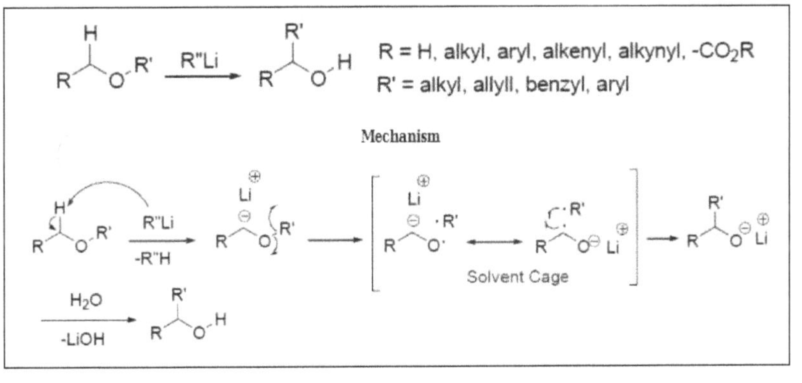

Aromatic Rearrangements

1. A number of rearrangements occur in aromatic compounds of the type:

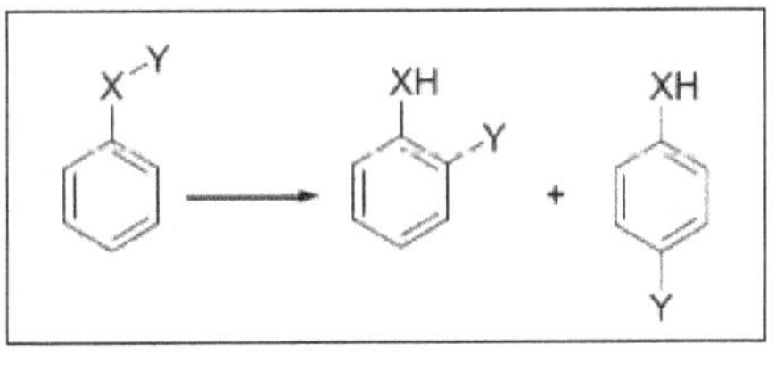

2. X is usually nitrogen or oxygen. Both intermolecular and intramolecular migrations are known.

3. These compounds undergo [5,5]-sigmatropic rearrangement in the presence of acid to give benzidines.

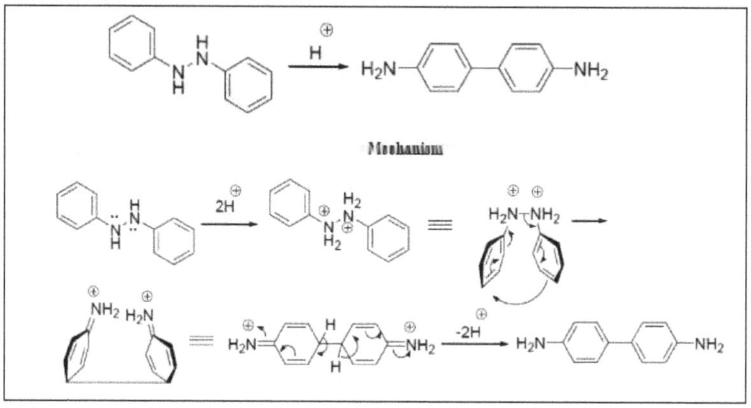

Claisen Rearrangement - Aryl allyl ethers undergo [3,3]-sigmatropic rearrangement on being heating to allylphenols.

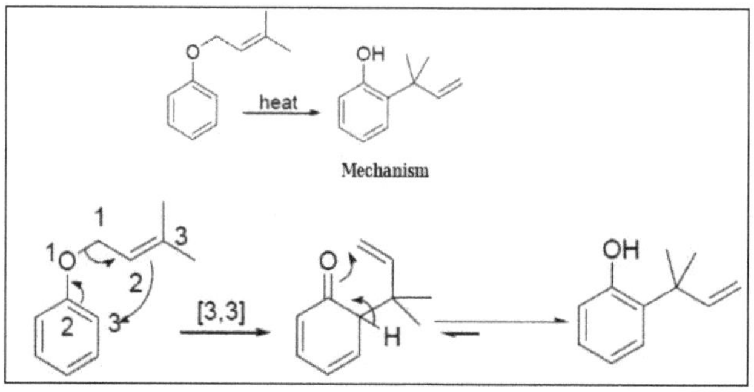

Named Organic Reactions

Claisen Rearrangement

The Claisen rearrangement is a powerful carbon–carbon bond-forming chemical reaction discovered by Rainer Ludwig Claisenin 1912. The heating of an allyl vinyl ether will initiate a [3,3]-sigmatropic rearrangement to give a γ,δ-unsaturated carbonyl.

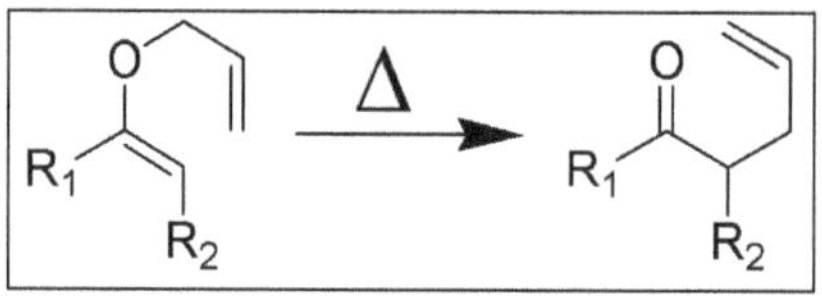

The Claisen rearrangement is the first recorded example of a-sigmatropic rearrangement.

The Claisen rearrangement is an exothermic, concerted (bond cleavage and recombination) pericyclic reaction. Woodward–Hoffmann rules show a suprafacial, stereospecific reaction pathway. The kinetics are of the first order and the whole transformation proceeds through a highly ordered cyclic transition state and is intramolecular. Crossover experiments eliminate the possibility of the rearrangement occurring via an intermolecular reaction mechanism and are consistent with an intramolecular process.

There are substantial solvent effects observed in the Claisen rearrangement, where polar solvents tend to accelerate the reaction to a greater extent. Hydrogen-bonding solvents gave the highest rate constants. For example, ethanol/water solvent mixtures give rate constants 10-fold higher than sulfolane. Trivalent organoaluminium reagents, such as trimethylaluminium, have been shown to accelerate this reaction.

The first reported Claisen rearrangement is the [3,3]-sigmatropic rearrangement of an allylphenyl ether to intermediate 1, which quickly tautomerizes to an ortho-substituted phenol.

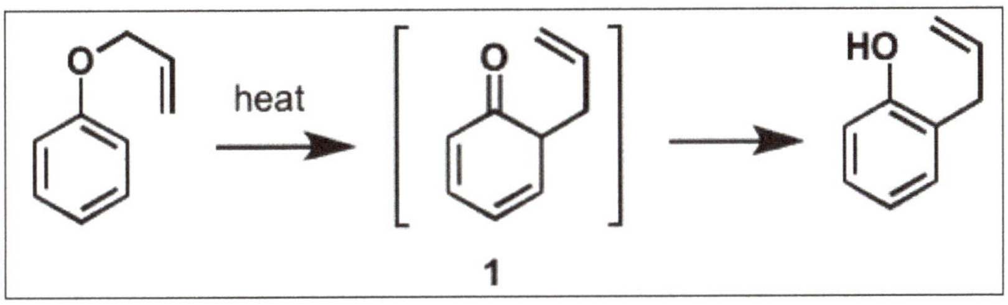

Meta-substitution affects the regioselectivity of this rearrangement. For example, electron withdrawing groups (e.g. bromide) at the meta-position direct the rearrangement to the ortho-position (71% ortho-product), while electron donating groups (e.g. methoxy), direct rearrangement to the para-position (69% para-product). Additionally, presence of ortho-substituents exclusively leads to para-substituted rearrangement products (tandem Claisen and Cope rearrangement).

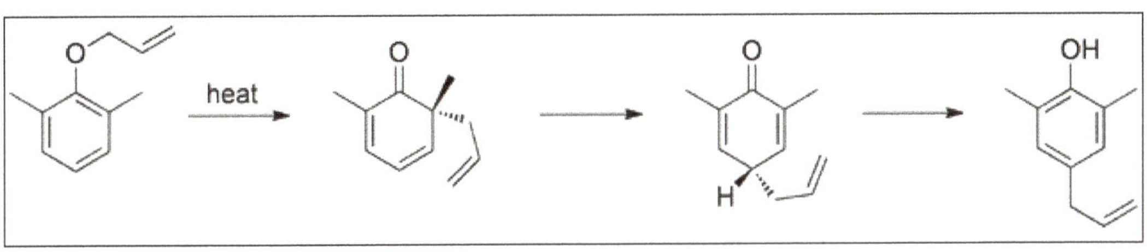

If an aldehyde or carboxylic acid occupies the ortho or para positions, the allyl side-chain displaces the group, releasing it as carbon monoxide or carbon dioxide, respectively.

Bellus–Claisen Rearrangement

The Bellus–Claisen rearrangement is the reaction of allylic ethers, amines, and thioethers with ketenes to give γ,δ-unsaturated esters, amides, and thioesters. This transformation was serendipitously observed by Bellus in 1979 through their synthesis of a key intermediate of an insecticide, pyrethroid. Halogen substituted ketenes (R1, R2) are often used in this reaction for their high electrophilicity. Numerous reductive methods for the removal of the resulting α-haloesters, amides and thioesters have been developed. The Bellus-Claisen offers synthetic chemists a unique opportunity for ring expansion strategies.

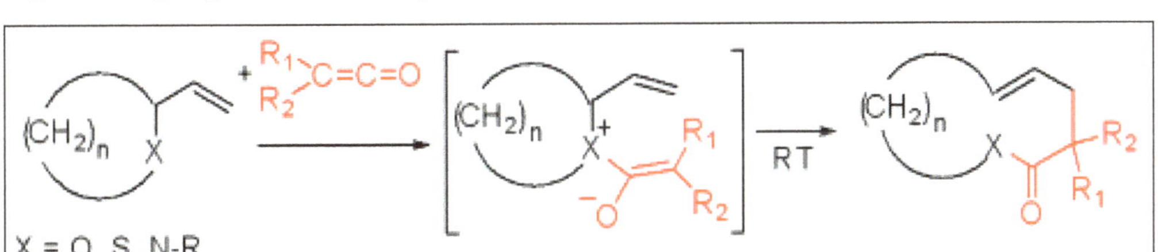

Eschenmoser–Claisen Rearrangement

The Eschenmoser–Claisen rearrangement proceeds by heating allylic alcohols in the presence of N,N-dimethylacetamide dimethyl acetal to form γ,δ-unsaturated amide. This was developed by Albert Eschenmoser in 1964. Eschenmoser-Claisen rearrangement was used as a key step in the total synthesis of morphine.

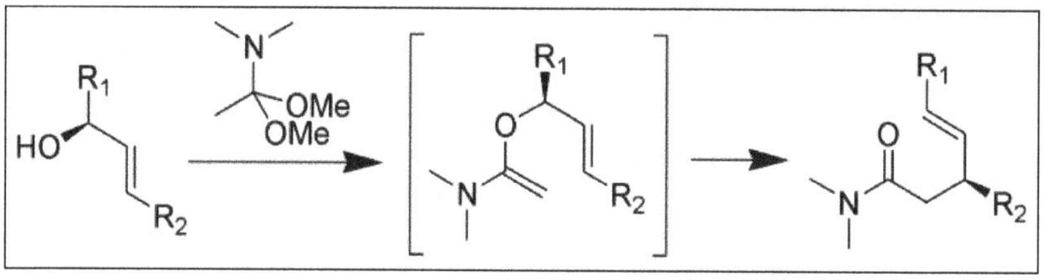

Ireland–Claisen Rearrangement

The Ireland–Claisen rearrangement is the reaction of an allylic carboxylate with a strong base (such as lithium diisopropylamide) to give a γ,δ-unsaturated carboxylic acid. The rearrangement proceeds via silylketene acetal, which is formed by trapping the lithium enolate with chlorotrimethylsilane. Like the Bellus-Claisen, Ireland-Claisen rearrangement can take place at room temperature and above. The E- and Z-configured silylketene acetals lead to anti and syn rearranged products, respectively. There are numerous examples of enantioselective Ireland-Claisen rearrangements found in literature to include chiral boron reagents and the use of chiral auxiliaries.

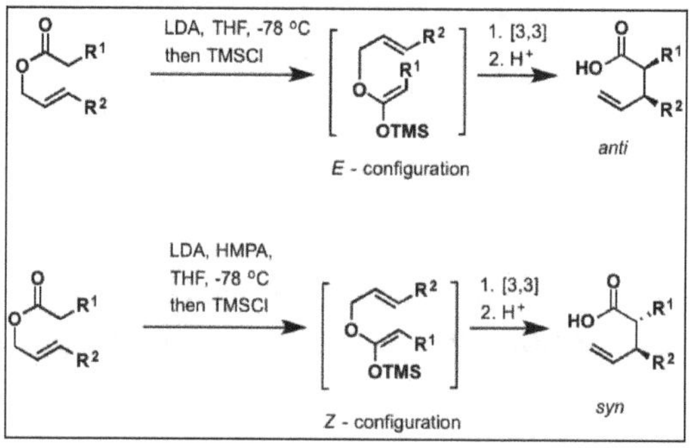

Johnson–Claisen Rearrangement

The Johnson–Claisen rearrangement is the reaction of an allylic alcohol with an orthoester to yield a γ,δ-unsaturated ester. Weak acids, such as propionic acid, have been used to catalyze this reaction. This rearrangement often requires high temperatures (100 to 200 °C) and can take anywhere from 10 to 120 hours to complete. However, microwave assisted heating in the presence of KSF-clay or propionic acid have demonstrated dramatic increases in reaction rate and yields.

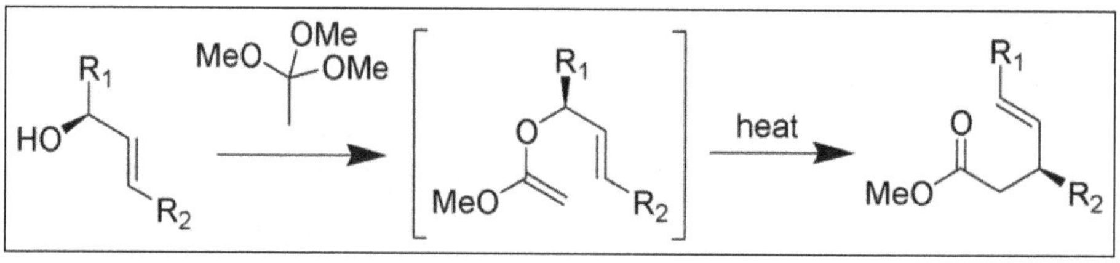

Mechanism:

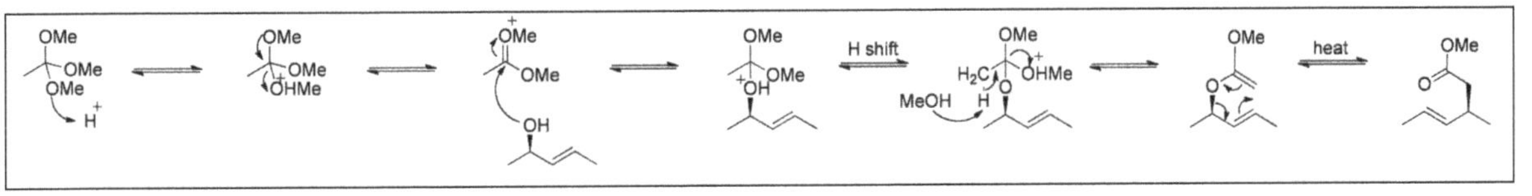

Photo-Claisen Rearrangement

The photo-claisen rearrangement is closely related to the photo-Fries rearrangement, that proceeds through a similar radical mechanism. Aryl ethers undergo the photo-Claisen rearrangement, while the photo-Fries rearrangement utilizes aryl esters.

Zwitterionic Claisen Rearrangement

Unlike typical Claisen rearrangements which require heating, zwitterionic Claisen rearrangements take place at or below room temperature. The acyl ammonium ions are highly selective for Z-enolates under mild conditions-

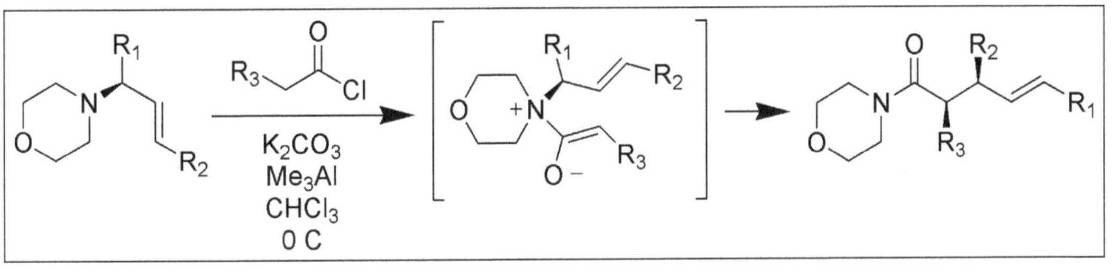

Claisen Rearrangement in Nature

The enzyme Chorismate mutase (EC 5.4.99.5) catalyzes the Claisen rearrangement of chorismate ion to prephenate ion, a key intermediate in the shikimic acid pathway (the biosynthetic pathway towards the synthesis of phenylalanine and tyrosine).

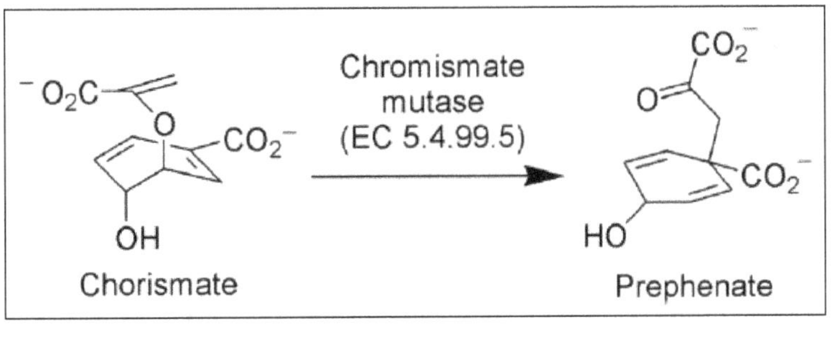

Sandmeyer Reaction

Sandmeyer reaction is a type of substitution reaction that is widely used in the production of aryl halides from aryl diazonium salts. Copper salts like chloride, bromide, or iodide ions are used as catalysts in this reaction. Notably, Sandmeyer reaction can be used to perform unique transformations on benzene. The transformations include hydroxylation, trifluoromethylation, cyanation, and halogenation.

Sandmeyer Reaction Mechanism

The Sandmeyer reaction follows a free radical mechanism. The reaction is actually a two-step process where the synthesis of aryl halides from primary aryl amines involves the formation of diazonium salts and the transformation of diazo intermediates into aryl halides (displacement with a nucleophile). Interestingly, the nucleophile can be a halide anion, cyanide, water, etc.

To elucidate further, the Sandmeyer reaction mechanism commences with a transfer of a single electron from the copper to the diazonium. This results in the formation of a non-participating diazo radical as well as copper(II) halide. A molecule of nitrogen gas is then released by the diazo radical to give aryl radical, which then reacts with the copper(II) halide to restore the catalyst [copper(I) halide]. After all this, the final product, aryl halide is obtained.

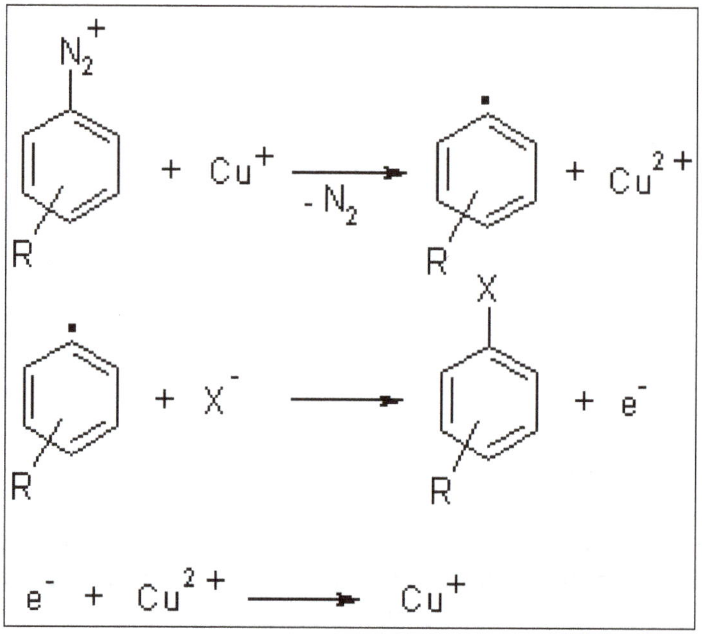

Formation of Nitrosonium Ion

We will look at Sandmeyer reaction mechanism during the formation of nitrosonium ion. In the experiment, we take sodium nitrite and acid to form nitrous acid. What follows next is the two protonation steps where one equivalent of water is removed. This results in the generation of the nitrosonium ion (nitrogen monoxide cation).

The ion acts as an electrophile when it is reacted further with an aromatic or heterocyclic amine (for e.g. aniline). Finally, the diazonium salt is formed. The reaction is carried out at temperatures ranging between 25-30 °C.

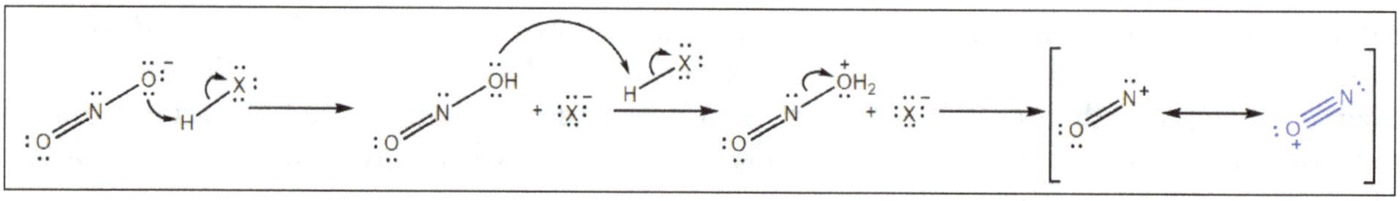

Benzenediazonium Ion Formation

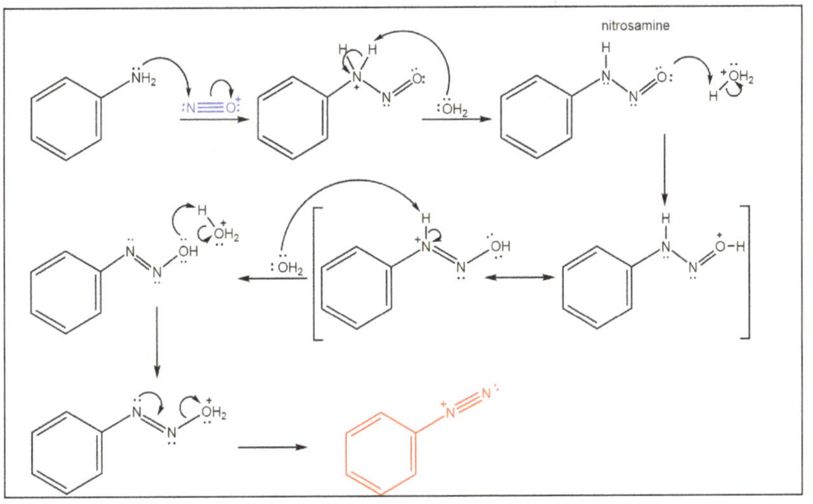

Example:

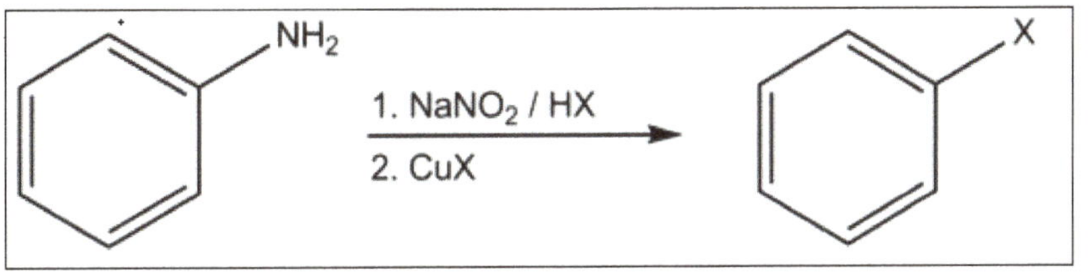

Importance of Sandmeyer Reaction

- The Sandmeyer reaction is very important when we are studying aromatic chemistry. The reaction can be used to generate substitution patterns that are sometimes not possible to achieve by direct substitution.
- In the reaction, a primary arylamine is mainly diazotized to create an aryl diazonium salt which is further reacted with a halide ion to produce the desired aryl halide product.
- It follows an electron-transfer-mechanism including free radicals.
- Sandmeyer reaction has wide synthetic applicability and is complementary to electrophilic aromatic substitution.

Applications of Sandmeyer Reaction

Today, a lot of variations of Sandmeyer reaction have been developed. As a result of these developments, the reaction has multiple synthetic applications. The different adaptations help in the synthesis of arylthioethers, aryl fluorides (the Schiemann reaction), phenols, and arylnitriles.

The common uses of the sandmeyer reaction are:

- It is used in hydroxylation to convert aryl amines to phenols leading to the formation of an aryl diazonium salt.

- To create aryl compounds during the process of trifluoromethylation. The compounds inherit unique chemical properties with a wide variety of practical applications.
- Sandmeyer reaction is used for the formation of benzonitriles through the process of cyanation. The reaction is also found during the synthesis of neoamphimedine, a chemical compound that targets topoisomerase II as an anti-cancer drug.
- To create aryl halides - Some of the solvents used in this process are diiodomethane to synthesize aryl iodides, bromoform to synthesize aryl bromides, chloroform to synthesize aryl chlorides.

Gattermann Reaction

The Gattermann reaction, (also known as the Gattermann aldehyde synthesis) is a chemical reaction in which aromatic compounds are formylated by hydrogen cyanide in the presence of a Friedel–Crafts catalyst (e.g. $AlCl_3$). It is named for the German chemist Ludwig Gattermann and is similar to the Friedel-Crafts reaction.

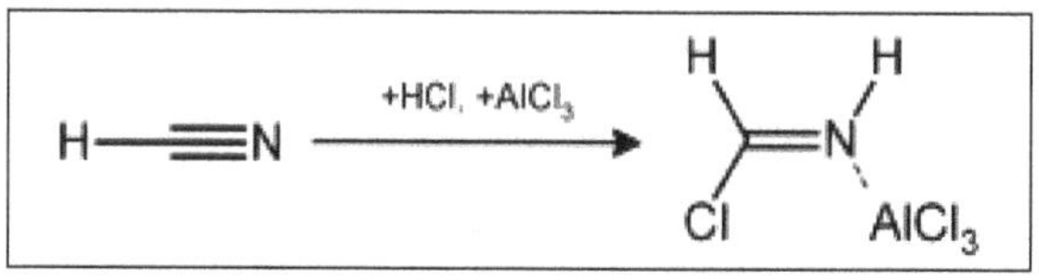

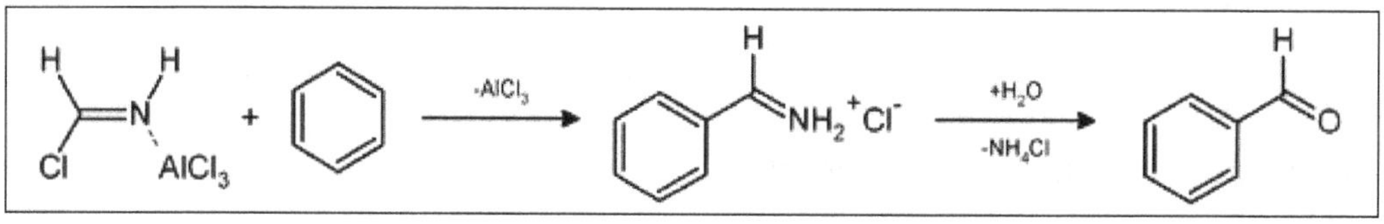

The reaction can be simplified by replacing the HCN/$AlCl_3$ combination with zinc cyanide. Although it is also highly toxic, $Zn(CN)_2$ is a solid, making it safer to work with than gaseous HCN; additionally, because the reaction uses HCl, $Zn(CN)_2$ also supplies the reaction with $ZnCl_2$ in-situ, where it acts as a Lewis acid catalyst. Examples of $Zn(CN)_2$ being used in this way include the synthesis of 2-Hydroxy-1-nafthaldehyde and Mesitaldehyde.

Gattermann–Koch Reaction

The Gattermann–Koch reaction, named after the German chemists Ludwig Gattermann and Julius Arnold Koch, refers to a Friedel–Crafts acylation reaction in which carbon monoxide, hydrochloric acid, and a Friedel–Crafts catalyst (e.g. $AlCl_3$) are used to produce aromatic aldehydes from various aromatic compounds, including derivatives of benzene and naphthalene:

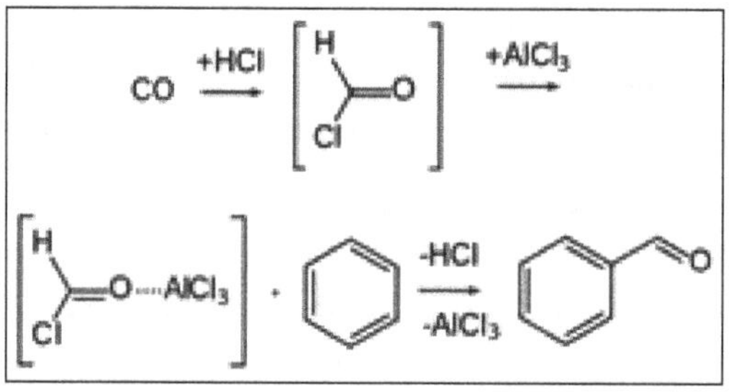

The applicability of the reaction includes many substituted aromatic derivatives, for example the conversion of toluene to p-tolualdehyde. However, unlike the Gattermann reaction with HCN, this reaction is not applicable to phenol and phenol ether substrates. Additionally, when Zinc chloride is used as the catalyst, the presence of traces of copper(I) chloride co-catalyst is often necessary.

Gattermann - Koch Reaction Mechanism

Gattermann Koch Reaction Mechanism begins with the formation of the reactive species with the help of the acid. The overall aim of the reaction is to attach a formyl group (-CHO group) to an aromatic system. An example of the Gattermann – Koch reaction is given below:

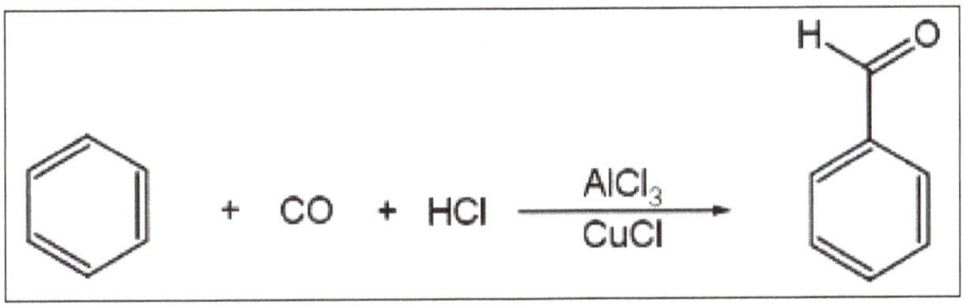

The Gattermann – Koch reaction is not applicable to phenol and phenol ether substrates. If zinc chloride is used as a catalyst in the Gattermann – Koch reaction, traces of copper(I) chloride is often necessary since it acts as a co-catalyst.

Step 1

The first step of the Gattermann Koch reaction mechanism is the generation of the reactive species which can later be used to react on the aromatic ring. Since carbon monoxide acts as a lewis base, it can accept a proton from the hydrochloric acid. This results in a positively charged molecule which has different resonance structures. One such resonance structure displays a positive charge on the carbon, explaining the reactivity of the hybrid. This species can act as an electrophile while reacting with the aromatic ring. However, it is more likely to be the target of a nucleophilic attack from the chloride ion in the hydrochloric acid.

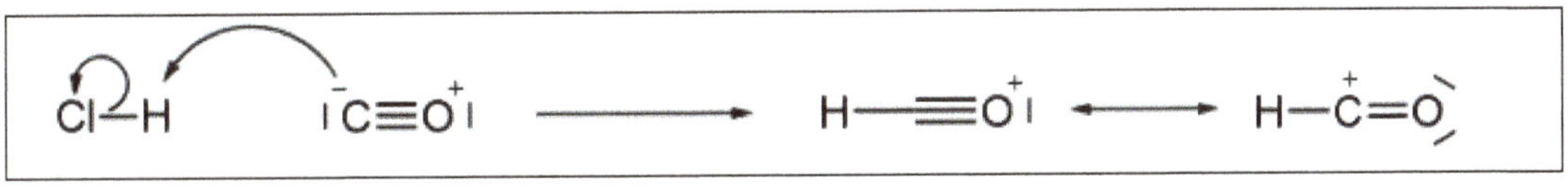

Step 2

When a Lewis acid (aluminium chloride) is added, it easily removes a chloride ion from the species. The species now reverts back to the reactive formyl cation.

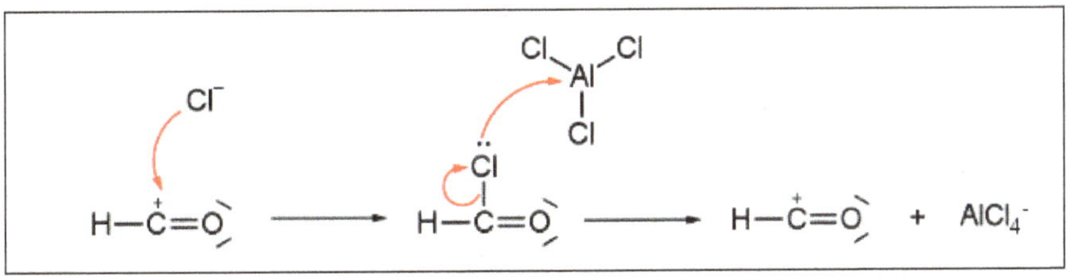

Step 3

An electrophilic aromatic substitution occurs at the aromatic ring. The aromatic ring acts as a nucleophile and donates an electron pair to the formyl cation. The temporary loss of aromaticity is quickly solved by the expulsion of a proton.

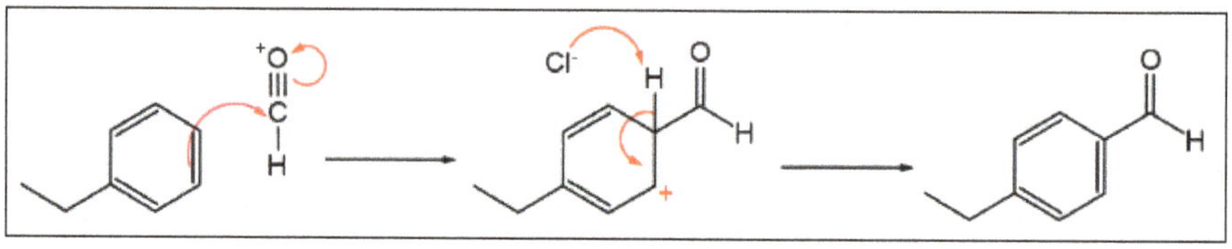

Thus, the formyl group is attached to the aromatic ring via the Gatterman – Koch reaction. In the example shown in the above mechanism, benzaldehyde is formed from the treatment of benzene with carbon monoxide and hydrochloric acid in the presence of aluminium chloride.

Wurtz Reaction

Wurtz reaction is an organic chemical coupling reaction wherein sodium metal is reacted with two alkyl halides in the environment provided by a solution of dry ether in order to form a higher alkane along with a compound containing sodium and the halogen.

The Wurtz reaction is a very useful reaction in the fields of organic chemistry and organometallic chemistry for the formation of alkanes. In this reaction, two different alkyl halides are coupled to yield a longer alkane chain with the help of sodium and dry ether solution.

This reaction is named after the French chemist Charles Adolphe Wurtz, who also discovered the aldol reaction. Apart from sodium, metals like silver, indium, activated copper, zinc, and iron can also be used in the Wurtz reaction in order to obtain alkanes.

The mechanism of this reaction involves free radicals, allowing for the possibility of side reactions that lead to the formation of alkenes as the product. A reaction related to the Wurtz Reaction in which aryl halides are used instead of alkyl halides is often called the Wurtz-Fittig reaction and is a very important named reaction in organic chemistry.

Wurtz Reaction Equation

The general form of the Wurtz reaction equation can be written as follows:

$$2R\text{-}X + 2Na \rightarrow R\text{-}R + 2Na^+ X^-$$

It can be observed from this equation that the two R groups are joined, yielding an alkane with a longer chain along with NaX, where X is a Halogen.

Wurtz Reaction Mechanism

The mechanism of the Wurtz reaction involves a free radical species denoted by R• which is a part of a halogen-metal exchange. This mechanism is somewhat similar to the formation of Grignard reagents. The carbon-carbon bond is formed in a nucleophilic substitution reaction in this reaction mechanism, which can be broken down into the following 3 steps:

Step1: The transfer of an electron from the metal (sodium in this case) to the halogen leads to the formation of an alkyl halide along with the metal halide. This reaction can be written as follows:

R-X + Na → R• + Na+R–

Step2: A different sodium atom now donates a single electron to the alkyl radical, leading to the formation of an alkyl anion as shown below:

R• + Na → R–Na+

Step3: The carbon belonging to the alkyl anion having a nucleophilic nature proceeds to displace the halogen in the alkyl halide via an SN2 reaction and form a covalent bond with the carbon which was bonded with the halogen. The reaction detailing this step is given below:

R–Na+ + R-X → R-R + Na+X–

The free radical mechanism for the Wurtz reaction involves the possibility of an alkene being produced as a side product. This side reaction is explained via the reaction provided below.

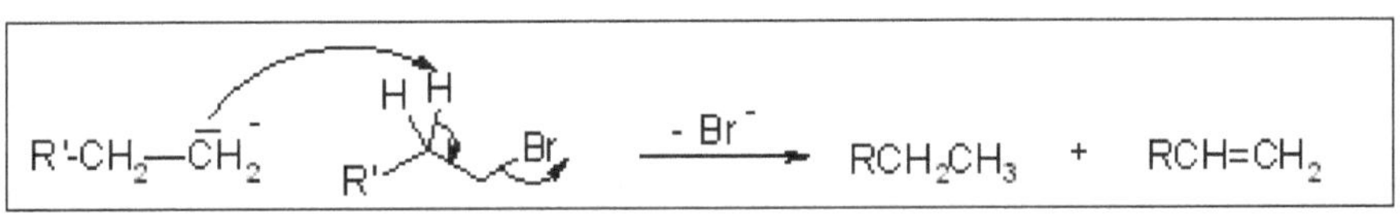

Thus, the required alkane product is formed in the Wurtz reaction mechanism. It can be noted that the reaction has relatively low yields due to the formation of multiple products.

Limitations of the Wurtz Reaction

A few limitations of this reaction are listed below:

- Commonly, only symmetric alkanes can be synthesized via this method since a mixture of alkane products are formed when dissimilar alkanes are reacted (these mixtures are difficult to separate).
- There exists a side reaction via which an alkene product is formed. Should the alkyl halides be bulky in nature, especially at the halogen end, there is a greater amount of alkene formed.
- Methane cannot be synthesized via the Wurtz reaction since the product of an organic coupling reaction must have at least two carbon atoms.
- The Wurtz coupling method generally fails when tertiary alkyl halides are used.

Fitting Reaction

In Fittig reaction, two aryl halides are coupled in presence of sodium metal in dry ether or tetrahydrofuran to furnish biaryls.

$$Ar - X + 2\,Na + X - Ar \xrightarrow{\text{Dry Ether}} Ar - Ar + 2\,NaX$$

Where Ar = aryl group, X = halogen

- The yields will be improved by using ultrasound, especially in two-phase reactions.
- A modification of reaction which involves, an alkyl halide and an aryl halide is called Wurtz-Fittig reaction.

The Wurtz-Fittig reaction mechanism can be explained either via the organo-alkali mechanism or the radical mechanism. The chemical reaction of aryl halides with alkyl halides and sodium metal with dry ether present to yield substituted aromatic compounds is called the Wurtz-Fittig reaction for the scientists – Charles Adolphe Wurtz and Wilhelm Rudolph Fittig (Wurtz for the discovery of the carbon-carbon bond formation from the coupling of two alkyl halides and Fittig for extending the approach to alkyl halide – aryl halide coupling). An example of such a reaction is shown below:

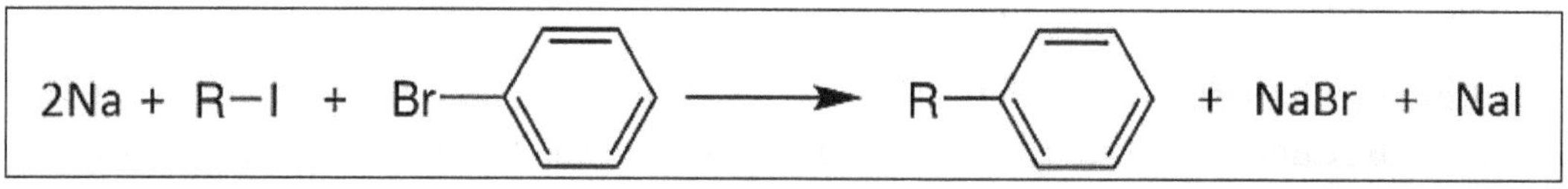

If the alkyl halide has more chemical reactivity than the aryl halide, it will form the carbon sodium bond first and act as a nucleophile to the aryl halide, helping the reaction in forming asymmetrical products. If the reactants are formed with halogens of different periods, the difference in reactivity between the alkyl halide and aryl halide can be achieved easily.

The Organo Alkali Mechanism

When the aryl halide is reacted with sodium metal, an intermediate organo-alkali compound is formed, which is followed by a nucleophilic attack of the alkyl halide as shown below. Thus, the required alkyl-aryl is formed.

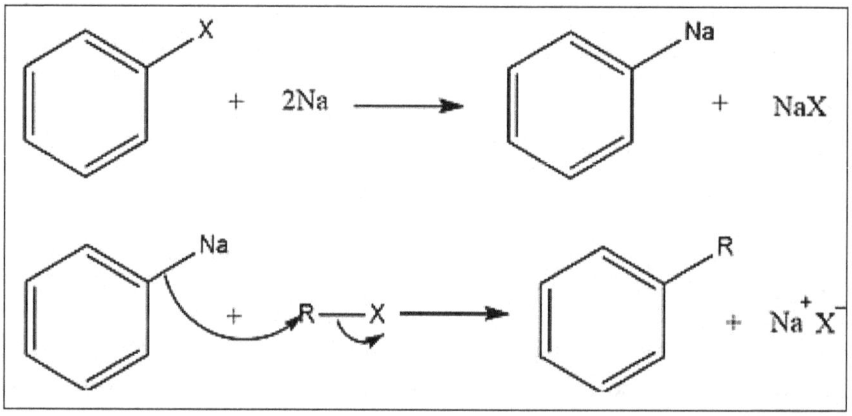

The Radical Mechanism

The sodium atom acts as a moderator for the formation of alkyl radicals and aryl radicals. These alkyl and aryl radicals now combine to form a substituted aromatic compound as shown below:

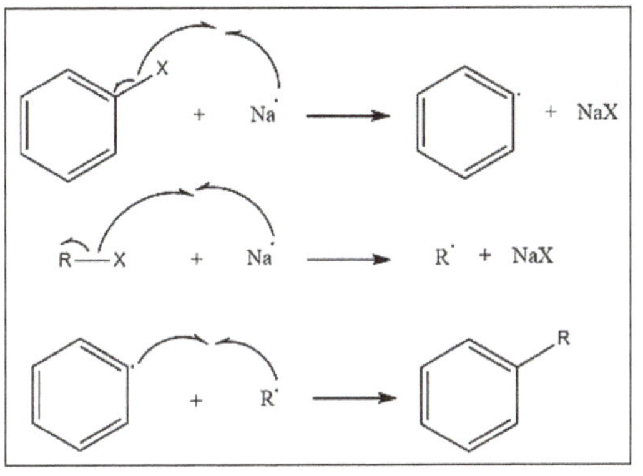

To conclude, the Wurtz-Fittig Reaction Mechanism can be described either via the radical mechanism or the organo-alkali mechanism, there exists empirical evidence for both. The reaction can also be conducted using metals other than sodium (such as potassium, iron, copper, and lithium). The reaction can also be called a coupling reaction. It requires an aprotic solvent as the reaction medium for which dry ether is quite suitable as it is a good non-polar aprotic solvent.

Reimer Tiemann Reaction

The Reimer Tiemann reaction is a chemical reaction named after chemists Karl Reimer and Ferdinand Tiemann. The reaction is used for the ortho-formylation of C_6H_5OH (phenols).

When phenols i.e. C_6H_5OH is treated with $CHCl_3$ (chloroform) in the presence of NaOH (sodium hydroxide), an aldehyde group (-CHO) is introduced at the ortho position of benzene ring leading to the formation of o-hydroxybenzaldehyde. The reaction is popularly known as Reimer Tiemann reaction.

A common example of Reimer Tiemann reaction is the conversion of phenol to salicylaldehyde (2-hydroxy benzaldehyde) as shown below:

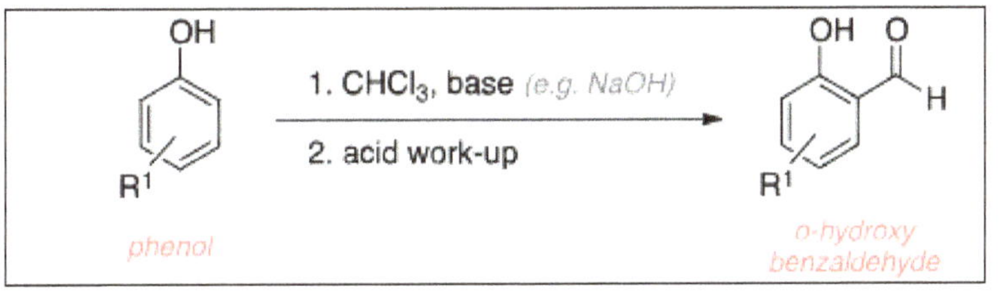

Since hydroxides are not readily soluble in chloroform, a biphasic solvent system is employed to carry out the reaction. This biphasic solvent system can consist of an aqueous hydroxide solution with an organic phase which contains the chloroform.

These two reagents that are separated are brought together for the reaction to occur. Techniques to bring these two reagents together include – rapid mixing, phase-transfer catalysts or the use of an emulsifying agent.

The reaction is quite effective when other hydroxy-aromatic compounds are used, naphthols for example. Heterocyclic organic compounds that are quite rich in electrons, such as pyrroles and indoles also can undergo the Reimer Tiemann reaction.

The reaction needs heat to initiate the process. However, once the reaction is begun, it can prove to be highly exothermic and further increase the reaction rate. This is the reason why the Reimer Tiemann reaction is prone to thermal runaways.

Reimer Tiemann Reaction Mechanism

The mechanism of Reimer Tiemann reaction begins with the deprotonation of chloroform by a strong base to form a chloroform carbanion. This chloroform carbanion quickly undergoes alpha elimination and gives rise to dichlorocarbene – the principle reactive species for this reaction.

The Reimer Tiemann reaction is an organic chemical reaction where phenol is converted into an ortho hydroxy benzaldehyde using chloroform, a base, and acid workup. This name reaction can also be described as the chemical reaction used for the ortho-formylation of phenols.

The mechanism of Reimer Tiemann Reaction can be explained in 6 major steps:

- The chloroform is deprotonated by the strongly basic aqueous hydroxide solution, giving the chloroform carbanion.
- This chloroform carbanion readily undergoes alpha elimination, giving dichlorocarbene as the product. As mentioned earlier, the dichlorocarbene is the principle reactive species.
- The aqueous hydroxide also deprotonates the phenol reactant, yielding a negatively charged phenoxide.
- This negative charge is now delocalized into the benzene ring, causing it to be far more nucleophilic.
- This results in a nucleophilic attack on the dichlorocarbene, forming an intermediate di-chloromethyl substituted phenol.
- This intermediate is subjected to basic hydrolysis to finally achieve the formation of the desired ortho-hydroxybenzaldehyde.

The illustration of the Reimer Tiemann reaction mechanism is given below:

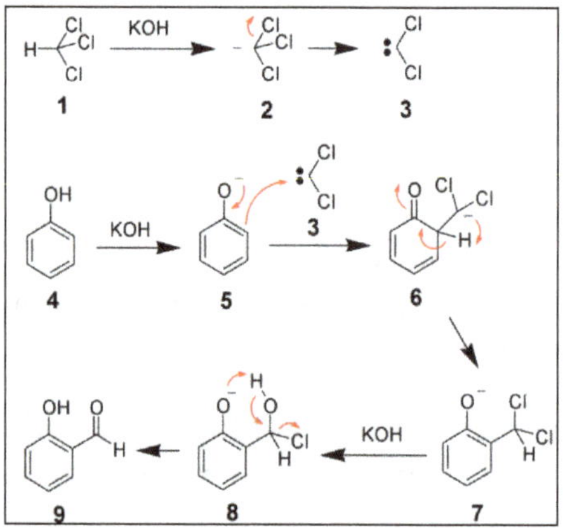

Thus, the given phenol is converted into an ortho-hydroxy benzaldehyde using chloroform, a base and acid workup. It can be noted that the carbene is highly electron-deficient due to the electron-withdrawing nature of its two chlorine groups. This is why it is strongly attracted to the phenoxide which is rich in electrons. The interaction favours ortho-formylation of a selective nature.

Clemmensen Reduction

Clemmensen reduction is an organic reaction which leads to the formation of alkanes by reduction of aldehydes or ketones in presence of acid and with amalgamated zinc. The Danish chemist discovered this reaction named Erik Christian Clemmensen. The substrate should not be acid sensitive, if the substrate is acid sensitive then reduction is carried out by Wolff-Kishner reduction.

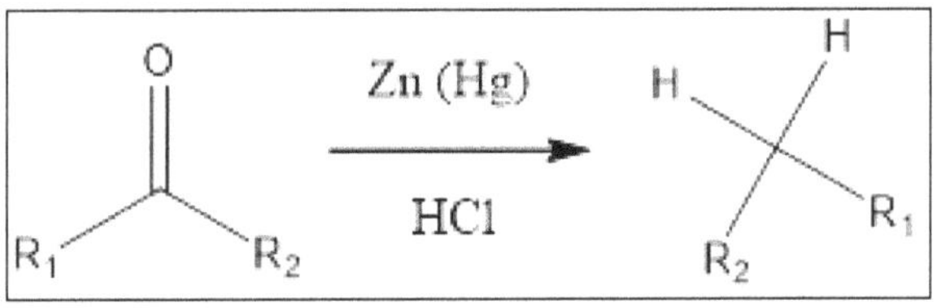

The mechanism of action is not exactly clear. There are two proposed mechanisms i.e. carbanionic mechanism where the attack of Zn is involved and carbenoid mechanism where Zn act as a surface to perform reaction and it is a radical process.

Carbanionic Mechanism

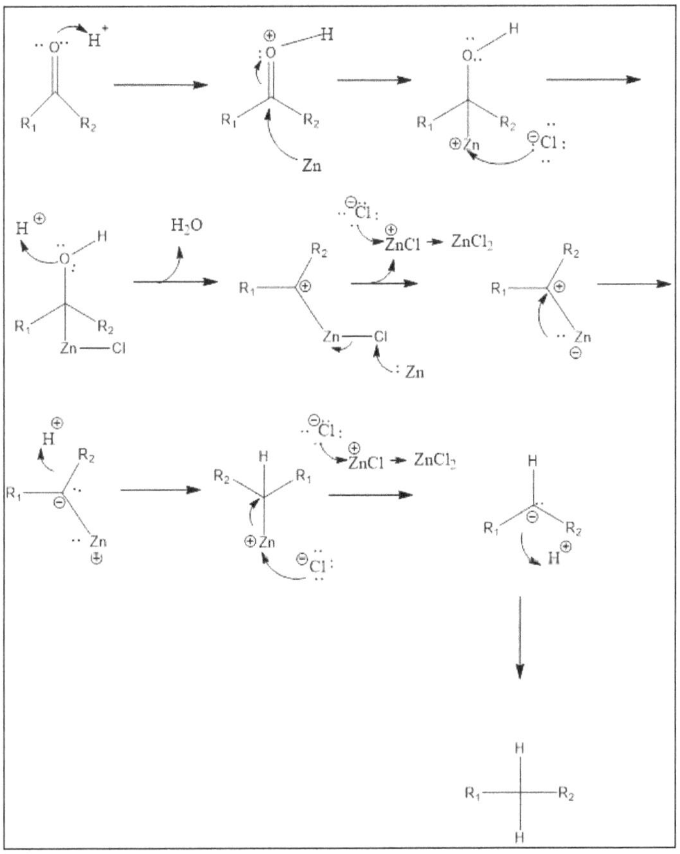

Carbenoid Mechanism

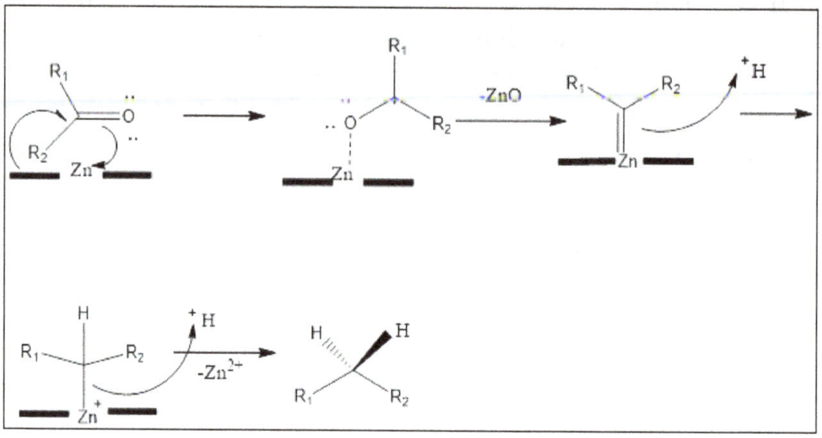

Wolff – Kishner Reduction

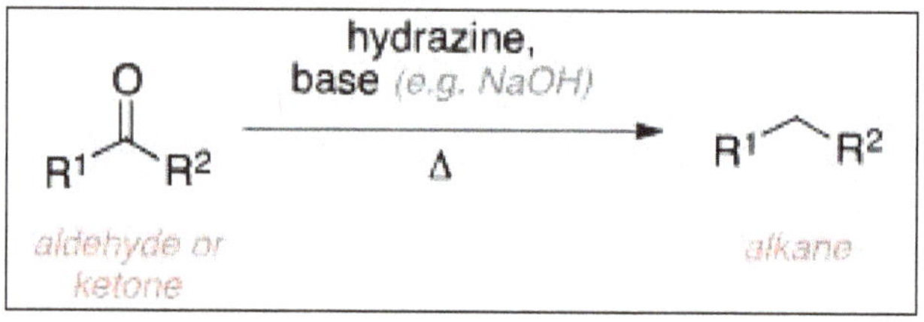

The Wolff-Kishner reduction is an organic reaction used to convert an aldehyde or ketone to an alkane using hydrazine, base, and thermal conditions. The mechanism begins with the attack of hydrazine of the aldehyde or ketone for form an imine. Proton transfer steps then result in the formation of a N=N bond. Deprotonation of the nitrogen and a rearrangement reaction result in the formation of a carbanion and the release of nitrogen gas. The carbanion then picks up a proton from water to regenerate the base catalyst and provides the final alkane product.

Reaction of adehydes or ketones with hydrazine produces a hydrazone:

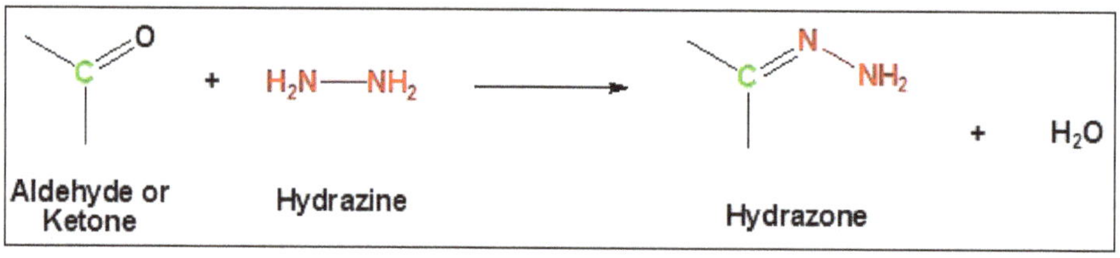

Reaction with a base and heat converts a hydrazone to an alkane:

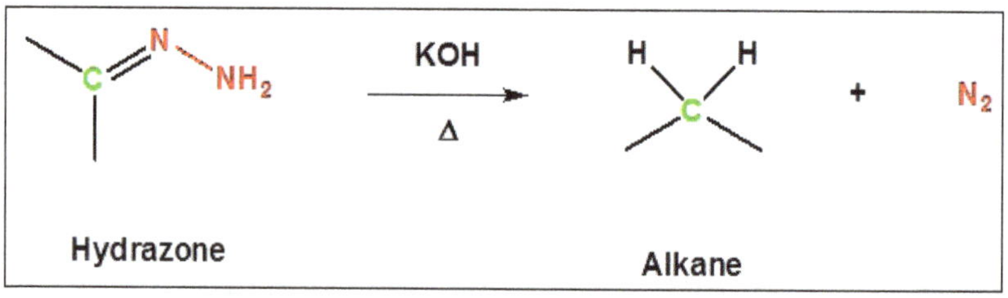

Both reactions together produces the Wolff-Kishner reduction:

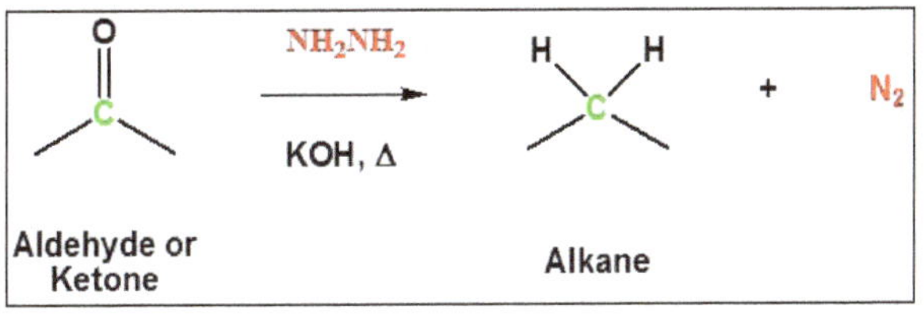

Example,

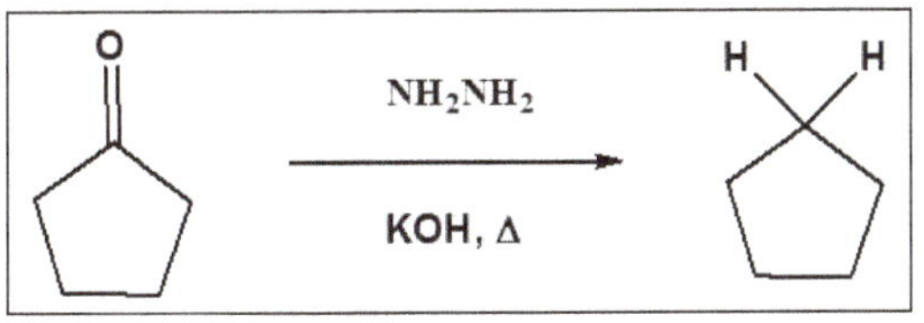

Mechanism of the Wolff-Kishner Reduction

1. Deprotonation of Nitrogen

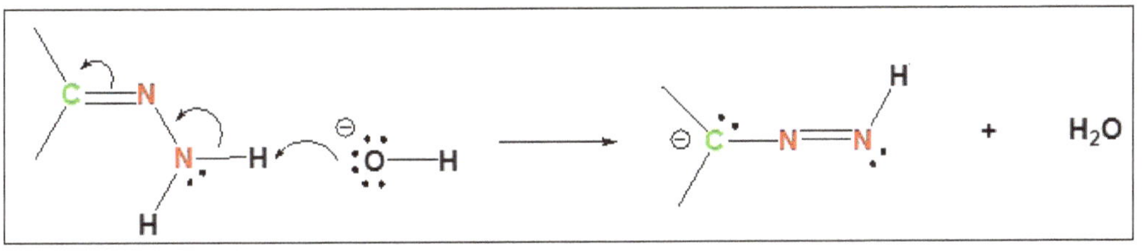

2. Protonation of the Carbon

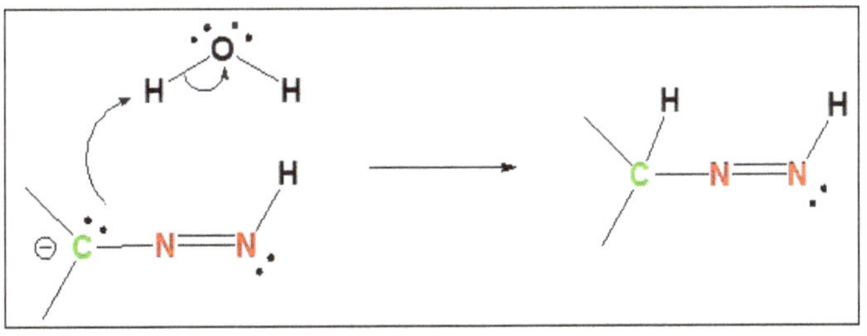

3. Deprotonation of Nitrogen

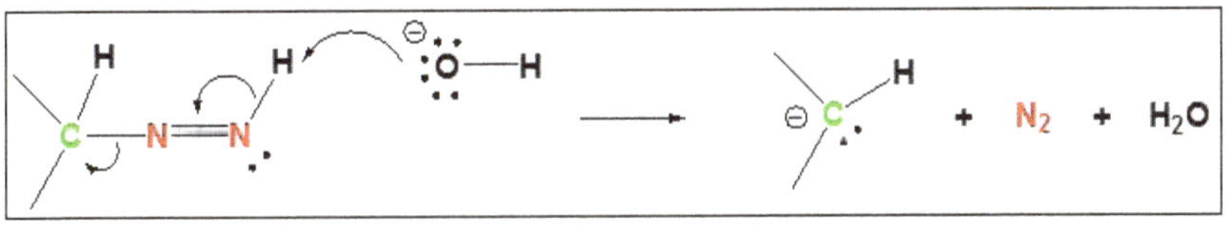

4. Protonation of Carbon

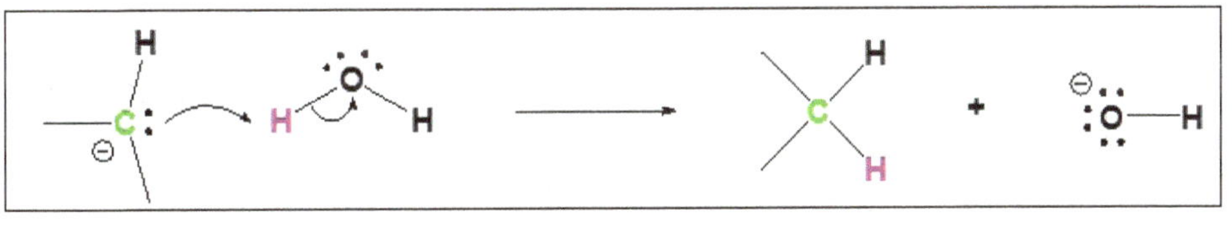

Aldol Condensation

An aldol condensation is a condensation reaction in organic chemistry in which an enol or an enolate ion reacts with a carbonyl compound to form a β-hydroxyaldehyde or β-hydroxyketone, followed by dehydration to give a conjugated enone.

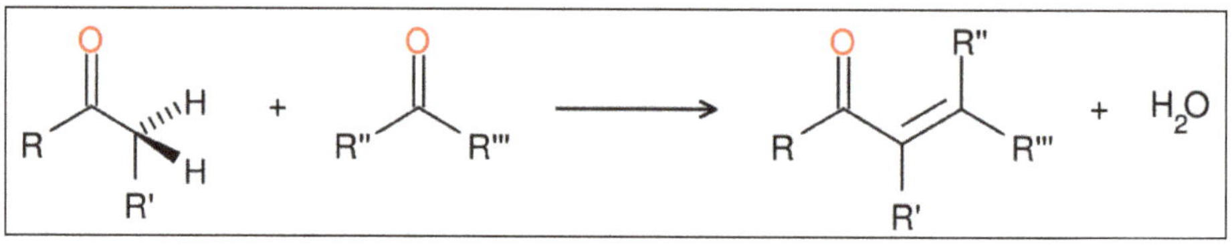

Aldol condensations are important in organic synthesis, because they provide a good way to form carbon–carbon bonds. For example, the Robinson annulation reaction sequence features an aldol condensation; the Wieland-Miescher ketone product is an important starting material for many organic syntheses. In its usual form, it involves the nucleophilic addition of a ketone enolate to an aldehyde to form a β-hydroxy ketone, or "aldol" (aldehyde + alcohol), a structural unit found in many naturally occurring molecules and pharmaceuticals.

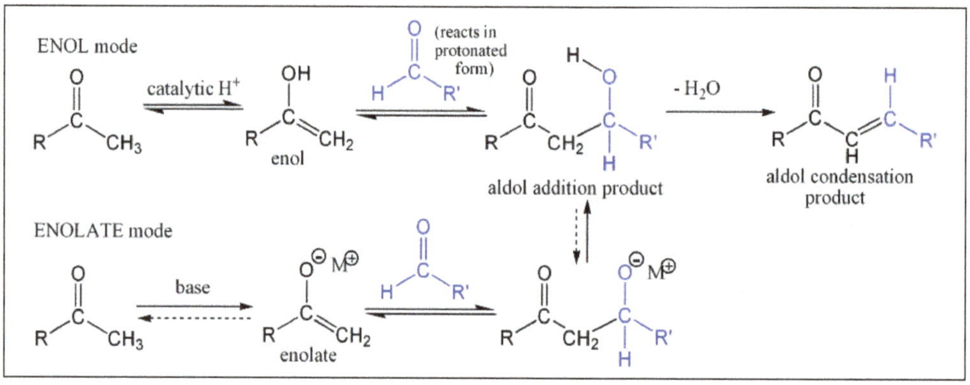

The name aldol condensation is also commonly used, especially in biochemistry, to refer to just the first (addition) stage of the process—the aldol reaction itself—as catalyzed by aldolases. However, the aldol reaction is not formally a condensation reaction because it does not involve the loss of a small molecule.

The reaction between an aldehyde/ketone and an aromatic carbonyl compound lacking an alpha-hydrogen (cross aldol condensation) is called the Claisen-Schmidt condensation. This reaction is named after two of its pioneering investigators Rainer Ludwig Claisen and J. G. Schmidt. An example is the synthesis of dibenzylideneacetone. Quantitative yields in Claisen-Schmidt reactions have been reported in the absence of solvent using sodium hydroxide as the base and plus benzaldehydes.

Mechanism

The first part of this reaction is an aldol reaction, the second part a dehydration—an elimination reaction (Involves removal of a water molecule or an alcohol molecule). Dehydration may be accompanied by decarboxylation when an activated carboxyl group is present. The aldol addition product can be dehydrated via two mechanisms; a strong base like potassium *t*-butoxide, potassium hydroxide or sodium hydride in an enolate mechanism, or in an acid-catalyzed enol mechanism. Depending on the nature of the desired product, the aldol condensation may be carried out under two broad types of conditions: kinetic control or thermodynamic control.

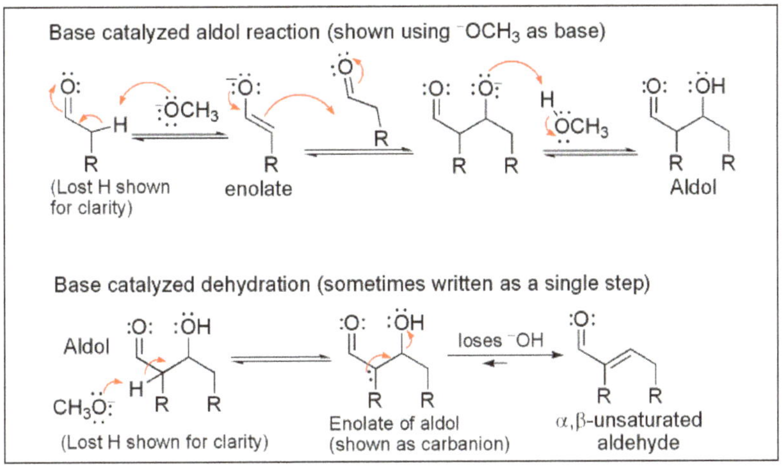

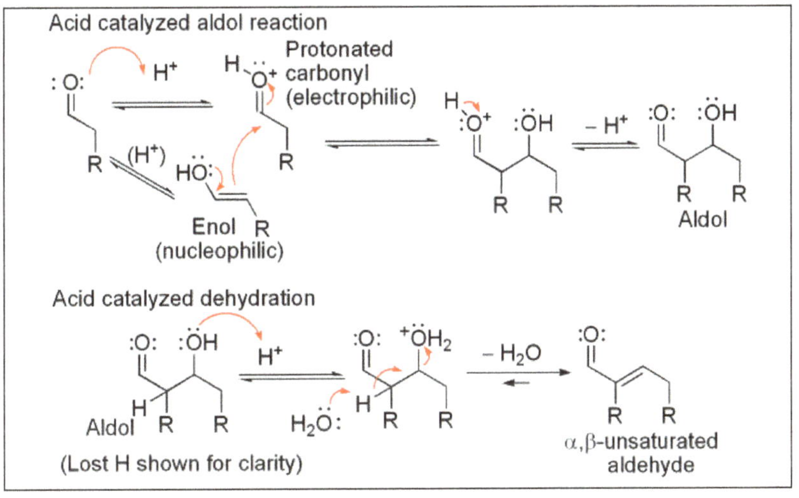

Condensation Types

It is important to distinguish the aldol condensation from other addition reactions of carbonyl compounds.

- When the base is an amine and the active hydrogen compound is sufficiently activated the reaction is called a Knoevenagel condensation.
- In a Perkin reaction the aldehyde is aromatic and the enolate generated from an anhydride.
- A Claisen condensation involves two ester compounds.
- A Dieckmann condensation involves two ester groups in the same molecule and yields a cyclic molecule.
- A Henry reaction involves an aldehyde and an aliphatic nitro compound.

- A Robinson annulation involves a α,β-unsaturated ketone and a carbonyl group, which first engage in a Michael reaction prior to the aldol condensation.
- In the Guerbet reaction, an aldehyde, formed *in situ* from an alcohol, self-condenses to the dimerized alcohol.
- In the Japp–Maitland condensation water is removed not by an elimination reaction but by a nucleophilic displacement.

Aldol Reactions under Equilibrating Conditions

To run successful aldol reactions under equilibrating conditions requires either that the donor and acceptor be the same, or that there be some molecular features that allow the donor and acceptor components to be distinguished. The self-aldolization of aldehydes can usually be stopped at the aldol stage, with more vigorous conditions (or a separate dehydration step) needed to form the enal.

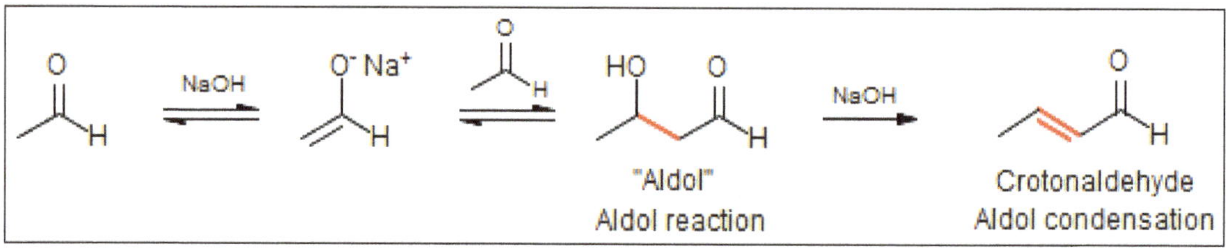

The self-aldolization of ketones is usually slow, and the equilibrium constant for aldol formation is often unfavorable. Only after dehydration does the reaction become thermodynamically favorable.

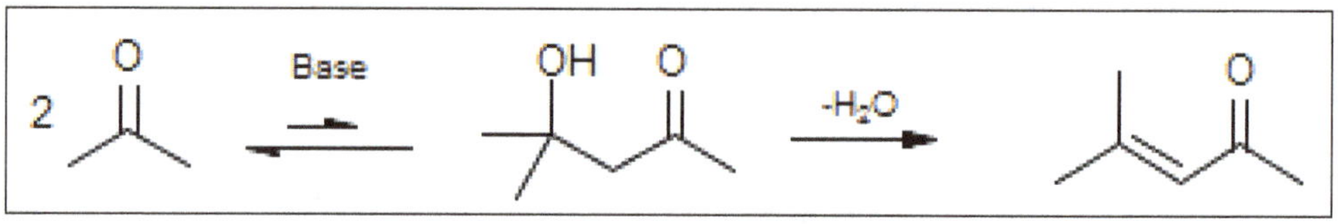

Crossed aldol reactions in which the acceptor cannot enolize are often successful. Some acceptors of this type are:

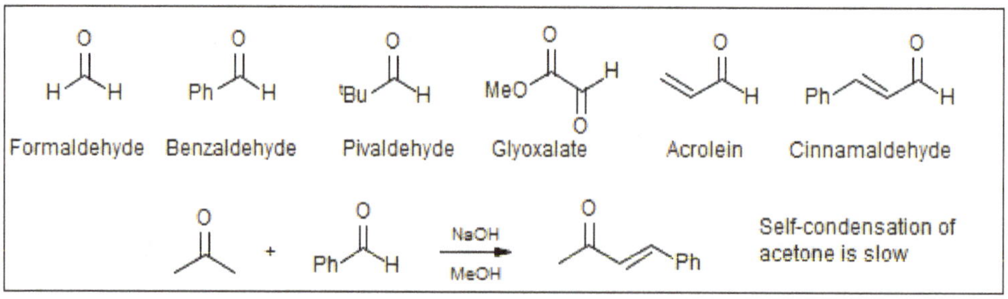

Cannizzaro Reaction

Cannizzaro reaction is a chemical reaction named after Stanislao Cannizzaro that involves the base-induced disproportionation of two molecules of a non-enolizable aldehyde to yield a carboxylic acid and a primary alcohol.

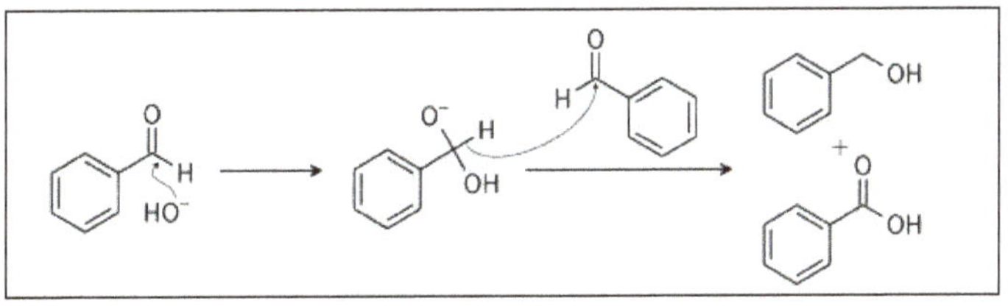

Cannizzaro Reaction Mechanism details the method to get one molecule of alcohol and one molecule of carboxylic acid from two molecules of a given aldehyde. Scientist Stanislao Cannizzaro, in 1853 succeeded in obtaining benzyl alcohol and potassium benzoate from benzaldehyde. The reaction is executed by a nucleophilic acyl substitution on an aldehyde where the leaving group attacks another aldehyde. A tetrahedral intermediate results from the attack of hydroxide on a carbonyl. This tetrahedral intermediate collapses, thereby reforming the carbonyl and transferring a hydride which attacks another colony.

Now, a proton is exchanged by acid and alkoxide ions. When a base of high concentration is introduced, the aldehyde forms an anion which has a charge of 2. From this, a hydride ion is transferred to a second molecule of the aldehyde, forming carboxylate and alkoxide ions. The alkoxide ion also obtains a proton from the solvent for the reaction.

Mechanism of Cannizzaro Reaction

Step 1

A nucleophile such as a hydroxide ion is used to attack the carbonyl group of the given aldehyde, causing a disproportionation reaction and giving rise to an anion carrying 2 negative charges.

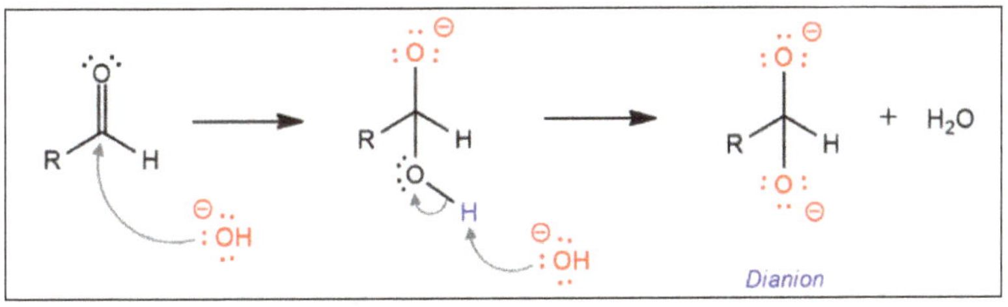

Step 2

This resulting intermediate can now function as a hydride reducing agent. Due to its unstable nature, the intermediate releases a hydride anion. This hydride anion proceeds to attack another aldehyde molecule. Now, the doubly charged anion is converted into a carboxylate anion and the aldehyde is converted into an alkoxide anion.

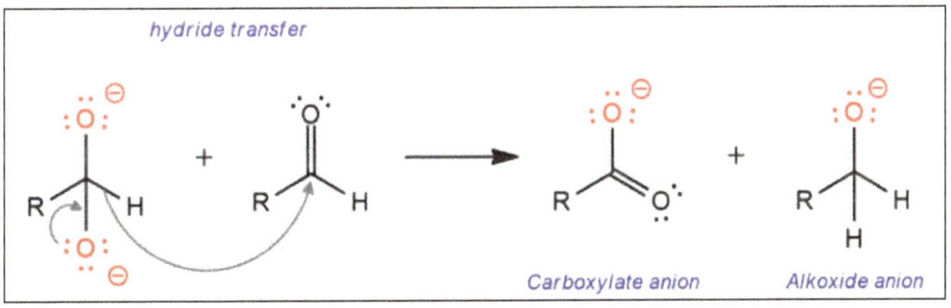

Step 3

In this final step, water offers a proton to the alkoxide anion which gives rise to the final alcohol product. The reaction can proceed since the alkoxide is more basic than water. Now, the carboxylate ion gives rise to the final carboxylic acid product when acid workup is used (the acid workup is required since carboxylate is less basic than water and therefore cannot obtain a proton from water).

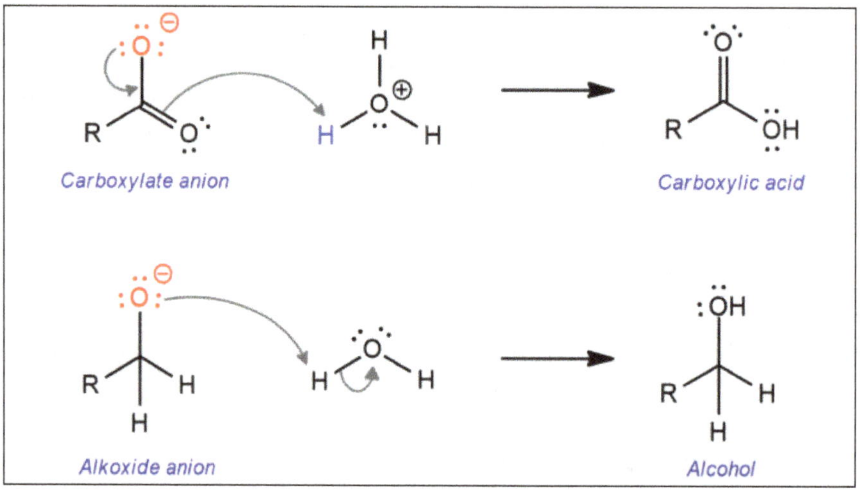

Crossed Cannizzaro Reaction

Aldehydes with alpha hydrogen atoms undergo deprotonation due to the strongly alkaline conditions of the reaction, leading to enolates and/or aldol reactions of these enolates where beta hydroxy aldehydes or ketones are obtained. Therefore, it is not surprising that the reaction produces only 50% of the required alcohol and carboxylic acid at ideal conditions. This is why the crossed Cannizzaro reaction is more commonly used. A sacrificial aldehyde is combined with a more valuable chemical and formaldehyde is used as a reductant, oxidizing it to sodium formate. The required alcohol is obtained from the reduction of the other aldehyde chemical. Since 2 different aldehydes can be completely converted into the required product, the yield of the valuable chemical is increased.

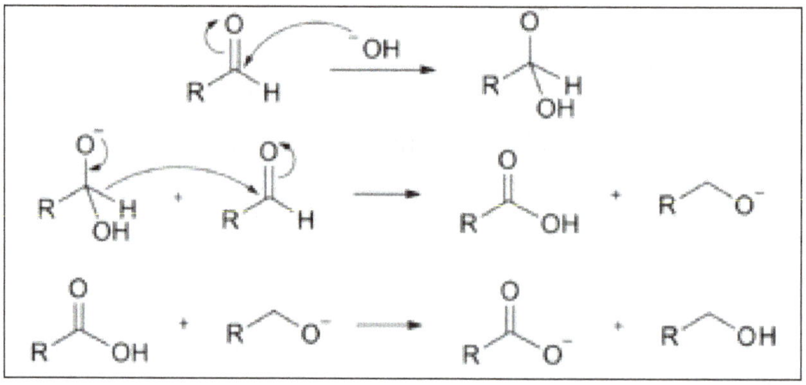

To conclude, the Cannizzaro reaction can be used to influence a disproportionation of a non-enolizable aldehyde. The cross Cannizzaro reaction is employed to increase the yield of the valuable chemical.

Gabriel Phthalimide Synthesis

The Gabriel synthesis is a method for making primary amines through an S_N2 reaction of a phthalimide with an alkyl halide followed by cleavage with hydrazine.

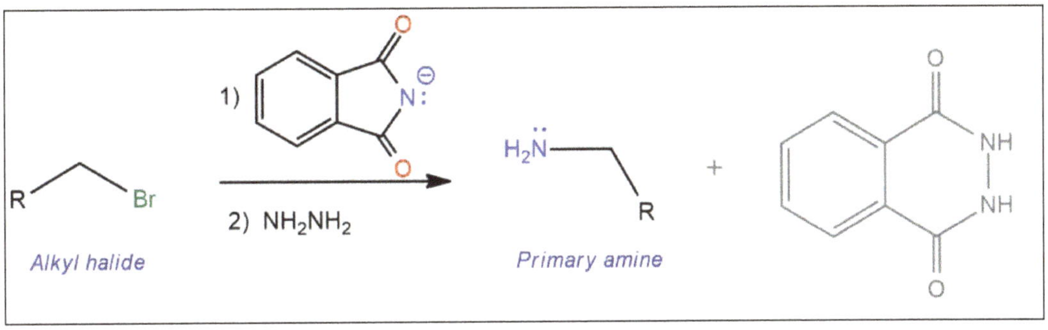

Gabriel Synthesis of Amines

The Gabriel synthesis is a nonreductive method for preparing primary amines with potassium phthalimide which is prepared by treating phthalimide with potassium hydroxide. The acidity of NH group is much greater than that of an ordinary amide because the nitrogen in the imide is adjacent to two carbonyl functions. Therefore, the deprotonation will be easily performed. The resulting anion function as a nucleophile and is readily alkylated by S_N2 reaction to form a carbon-nitrogen bond. This reaction works best with primary alkyl halides and sometimes with secondary, but tertiary cannot be used. Then this product can be hydrolyzed by acid or base.

Hydrolysis of Amides under Acidic Conditions

Acidic conditions are more common than basic conditions. An ammonium ion is produced which must be treated with a base to remove the charge and formed an amine.

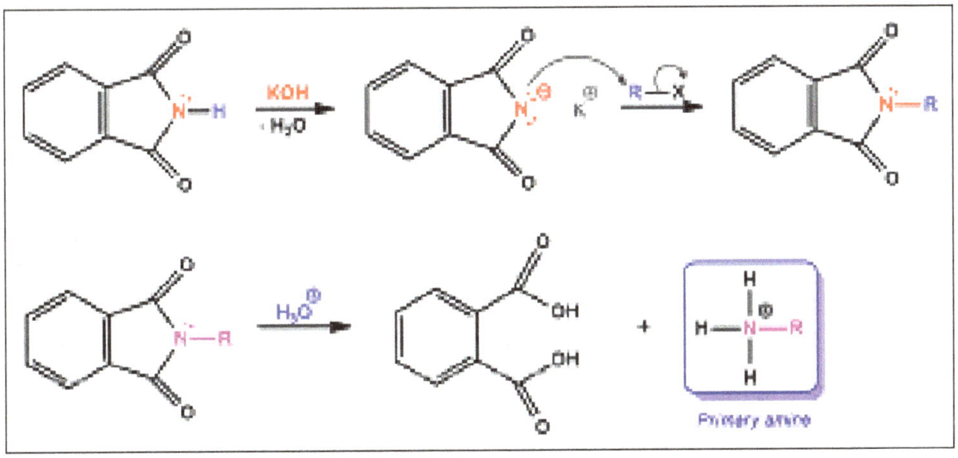

Hydrolysis of Amides under Basic Conditions

This is one of many predicted mechanisms. Hydrazine functions as a nucleophile and attacks both carbonyl functions with both nitrogen atoms. Primary amine and phthalimide hydrazide are obtained as products of this mechanism.

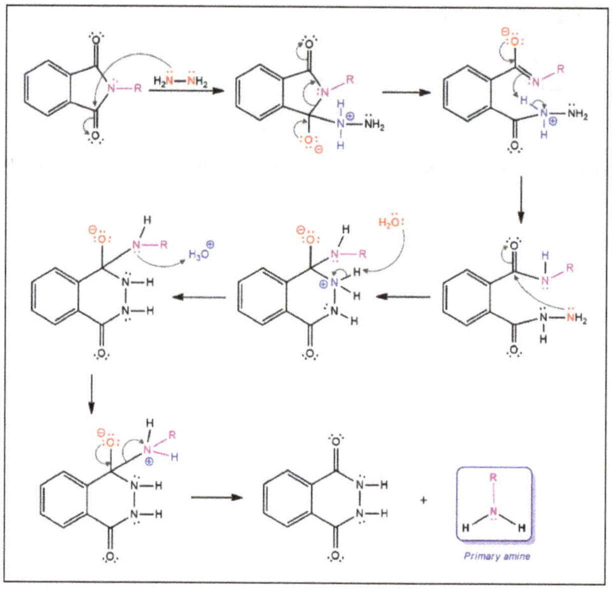

Mechanism of Gabriel Synthesis

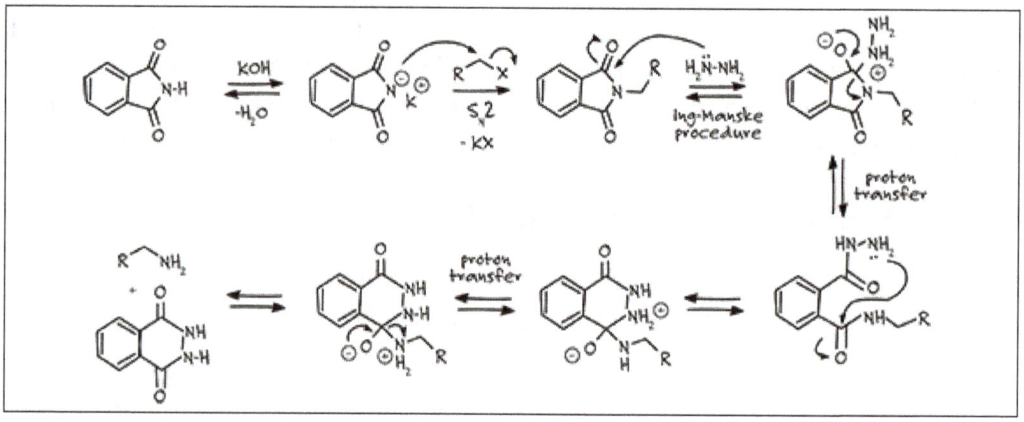

Kolbe's Reaction

Kolbe reaction, also known as Kolbe Schmitt Reaction is a type of addition reaction named after Hermann Kolbe and Rudolf Schmitt. When phenol is treated with sodium hydroxide, phenoxide ion is generated. The phenoxide ion generated is more reactive than phenol towards electrophilic aromatic substitution reaction. Hence, it undergoes electrophilic substitution reaction with carbon dioxide, which is a weak electrophile. Ortho-hydroxybenzoic acid (salicylic acid) is formed as the primary product. This reaction is popularly known as Kolbe's reaction.

An example of the Kolbe reaction (Kolbe Schmitt Reaction) is given below:

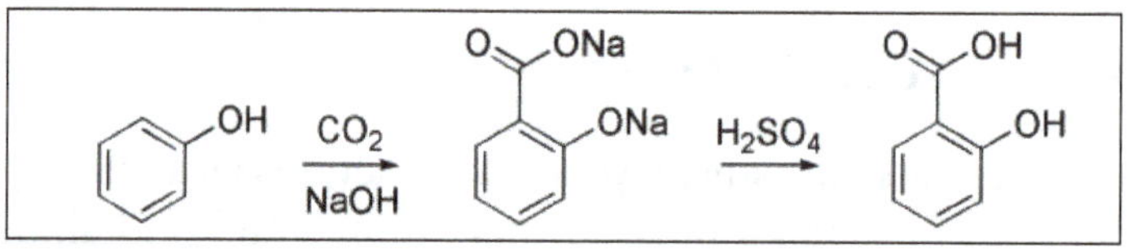

Kolbe's Reaction – Treating phenol with sodium hydroxide generates phenoxide ion.

The mechanism of Kolbe reaction proceeds through the nucleophilic addition of phenoxide to carbon dioxide, yielding the salicylate.

The salicylate formed further reacts with the acid to form salicylic acid. It is a carboxylation reaction where sodium phenoxide is heated with carbon dioxide under a pressure of 100 atmospheres and a temperature of 125 degrees Celsius, and the resulting product is treated with sulfuric acid to yield salicylic acid (an aromatic hydroxy acid).

Applications of Kolbe Reaction

- When Potassium Hydroxide is used in the Kolbe reaction, 4-Hydroxybenzoic acid can be accessed. This is an important precursor for parabens (parahydroxybenzoate or ester of para-hydroxy benzoic acid, used as a biocide in cosmetic products).
- Kolbe reaction can also be used for the industrial synthesis of 3-hydroxy-2-naphthoic acid, which is a common precursor to azo dyes and pigments.
- The salicylic acid can be used to make aspirin by reacting it with acetic anhydride. Aspirin is commonly used as a painkiller.

Kolbe Reaction Mechanism

The Kolbe reaction can be classified as a carboxylation chemical reaction. The reaction occurs when sodium phenoxide is allowed to absorb carbon dioxide and the resulting product is heated at a temperature of a 125-degree celsius and a pressure of over a hundred atmospheres. An unstable intermediate is now formed.

This unstable intermediate goes through a proton shift, leading to the formation of sodium salicylate. Now, this mixture is treated with sulfuric acid. The acidification of the mixture yields the salicylic acid. The illustration for the Kolbe reaction mechanism is given below:

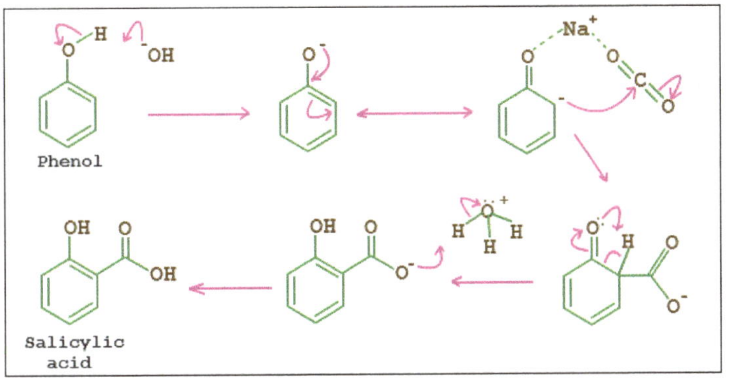

Thus, the required aromatic hydroxy acid – salicylic acid is produced via the Kolbe reaction. It can be observed that there is a nucleophilic addition of sodium phenoxide to carbon dioxide gas to form the salicylate in the mechanism.

Corey-House-Posner-Whitesides Reaction

The Corey–House synthesis (also called the Corey–Posner–Whitesides–House reaction and other permutations) is an organic reaction that involves the reaction of a lithium diorganylcuprate (R_2CuLi) with an organyl (pseudo)halide (R′X) to form a new alkane, as well as an ill-defined organocopper species and lithium halide as byproducts.

$$Li^+[R-Cu-R]^- + R'-X \rightarrow R-R' + "RCu" + Li^+X^-$$

In principle, a carbanion equivalent such as an organolithium or Grignard reagent can react directly (without copper) with an alkyl halide in a nucleophilic substitution reaction to form a new carbon–carbon bond. However, aside from the use of metal acetylides as nucleophiles, such a process rarely works well in practice due to metal–halogen exchange and/or the formation of large amounts of reduction or elimination side-products; as a solution to this problem, the Corey–House reaction constitutes a general and high yielding method for the joining of two alkyl groups or an alkyl group and an aryl group.

Reaction Process and Mechanism

The Corey-House synthesis is preceded by two preliminary steps to prepare the requisite Gilman reagent from an alkyl halide. In the first step, the alkyl halide is treated with lithium metal in dry ether to prepare an alkyllithium reagent, RLi; the starting alkyl halide for the lithiation step can be a primary, secondary or tertiary alkyl chloride, bromide, or iodide:

$$R-X + 2Li \rightarrow RLi + Li^+X^-$$

In the second step, a lithium dialkylcuprate, also known as a Gilman reagent is prepared from the alkyllithium by treatment with cuprous iodide (CuI) in a transmetalation reaction:

$$2RLi + CuI \rightarrow Li^+[R-Cu-R]^- + LiI$$

If the use of alkyllithium reagents is precluded by functional group incompatibility, transmetalation from other metals (e.g., Mg, Zn, Al, B) may be considered as alternatives for the preparation of the organocopper reagent.

The Corey-House synthesis process is the reaction between the organocopper reagent, usually a lithium dialkylcuprate as prepared above, and a second alkyl (pseudo)halide or an aryl iodide; this results in the formation of a C–C bond between the two organic fragments:

$$Li^+[R-Cu-R]^- + R'-X \rightarrow R-R' + "RCu" + LiX$$

From the stoichiometry, it is apparent that one equivalent of the R group is wasted as an ill-characterized alkylcopper species (likely polymeric; usually converted to RH upon aqueous workup) in the most common form of the Corey–House synthesis. To avoid this for cases where R is a precious or complex fragment, a reagent $(R)(R_U)CuM$, where R_U is an untransferable dummy ligand (e.g., R_U = cyano, alkynyl, 2-thienyl, etc.) can be prepared and used instead.

It is important to note that when R and R′ are different, only the cross product R–R′ is obtained; R–R or R′–R′ are not formed in significant amounts; the Corey–House reaction is therefore an example of a cross-coupling reaction. The Corey–House synthesis is, in fact, one of the earliest transition metal-mediated cross-coupling reactions to be discovered.

In the case of alkyl bromides and tosylates, inversion of configuration is observed when an configurationally pure alkyl electrophile is used; the reaction is believed to proceed via an S_N2-like mechanism to give a copper(III) species, which undergoes reductive elimination to give the coupling product. When alkyl iodides are used, scrambling of configuration is observed, and cyclization products are observed to form for alkyl iodides with an olefin tether, both of which are indicative of the involvement of radicals.

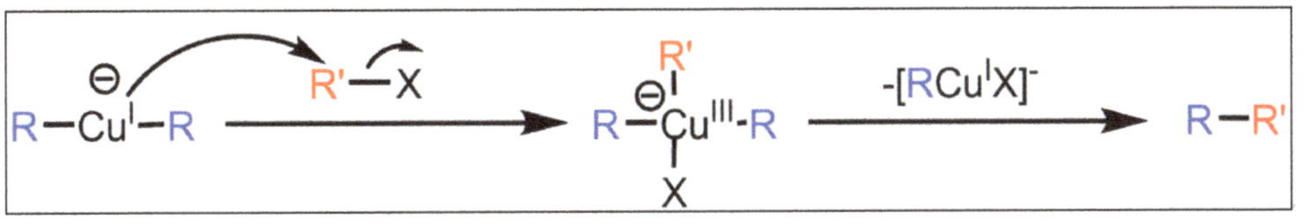

It is important to note that for this reaction to work successfully, the alkyl (pseudo)halide coupling partner should be methyl, benzylic, allylic, 1° alkyl, or 2° cycloalkyl. In most cases, 3° and acyclic 2° electrophiles give unsatisfactory results. On the other hand, sterically hindered organocopper reagents, including 3° and other branched alkyl reagents, are generally tolerated. However, aryl bromides, iodides and sulfonates, which do not ordinarily undergo nucleophilic substitution in the absence of a transition metal, can be used successfully as coupling partners.

Catalytic Version

In 1971, Kochi reported that Grignard reagents and alkyl bromides could be coupled using a catalytic amount of lithium tetrachlorocuprate(II), a process that was extended to alkyl tosylates by Schlosser and Fouquet. In the catalytic process, the Grignard reagent undergoes transmetalation with the copper salt or complex to generate an organocuprate as a catalytic intermediate, which then undergoes reaction with the (pseudo)halide electrophile to form the coupling product and release the copper and complete the catalytic cycle.

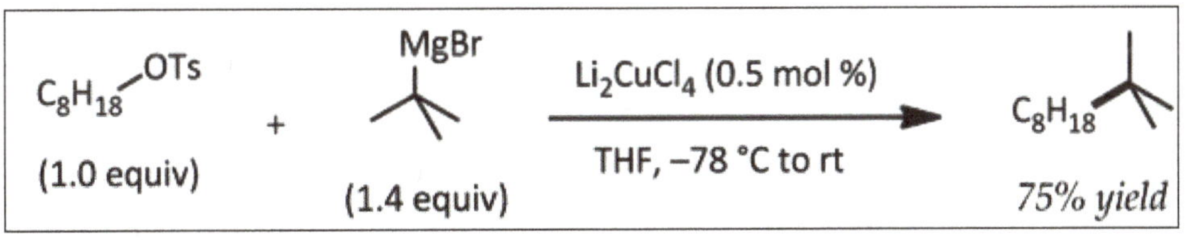

Under recently discovered conditions, using TMEDA as the ligand for copper and lithium methoxide as a base additive, it is now possible to couple 1°, 2°, and 3° Grignard reagents with 1° and 2° alkyl bromides and tosylates in high yields with nearly exclusive stereoinversion. Even β-branched 2° alkyl tosylates react to give coupling product in moderate yield, greatly expanding the scope of the catalytic Corey–House synthesis (Kochi–Schlosser coupling).

Grignard Synthesis

The Grignard reaction involves the synthesis of alcohol compounds from an aldehyde or ketone, using Grignard reagents. An alcohol group is one that has the molecular formula of ROH. They have hydroxide, OH, molecules attached to any molecule.

We can think of aldehydes and ketones as siblings. Both of them have parents called a carbonyl group, which contains a carbon atom double bonded to an oxygen atom.

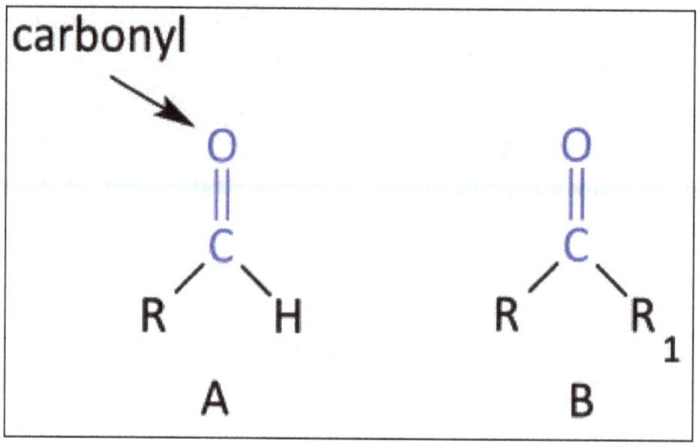

Molecular structure of an (A) Aldehyde and (B) Ketone

If we go back to our definition of a Grignard reaction we see that a very special reagent is needed, called the Grignard reagent. In general, reagents are very helpful during chemical experiments. They help ensure a reaction takes place. The specific Grignard reagent is formed when a metal atom attaches to an alkyl halide. Alkyl halides are organic groups that contain alkyl groups and a halogen, which we can spot on Group 17 of the periodic table.

One way to describe a Grignard reagent is the term 'organometallic compound.' Organometallic compounds are formed when a metal is attached to the carbon atom in a molecule. When working with the Grignard reagent, you will often use the magnesium metal. These compounds, containing magnesium, are called organomagnesium compounds.

So why is the Grignard reaction so useful? Well, it all boils down to the formation of carbon-carbon bonds. Grignard created a reaction that not only yields a wide variety of alcohol products, but also - and more importantly - provides scientists with an easy way to make new carbon-carbon bonds. Depending on whether you use an aldehyde or ketone, the type of alcohol you make may vary.

Mechanism of the Grignard Reaction

While the reaction is generally thought to proceed through a nucleophilic addition mechanism, sterically hindered substrates may react according to an SET (single electron transfer) mechanism:

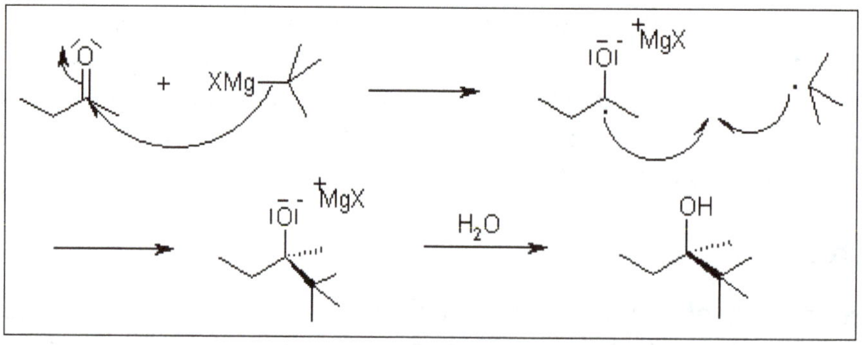

With sterically hindered ketones the following side products are received:

The Grignard reagent can act as base, with deprotonation yielding an enolate intermediate. After work up, the starting ketone is recovered.

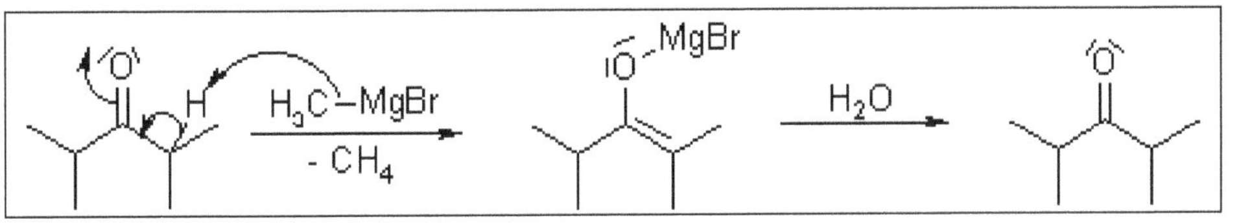

A reduction can also take place, in which a hydride is delivered from the β-carbon of the Grignard reagent to the carbonyl carbon via a cyclic six-membered transition state.

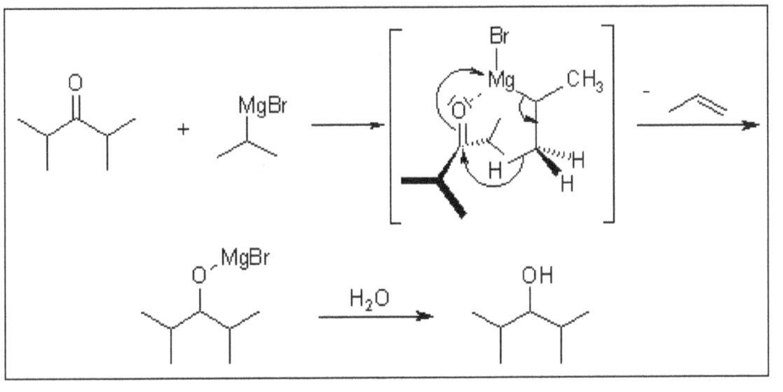

Additional Reactions of Grignard Reagents

With carboxylic acid chlorides:

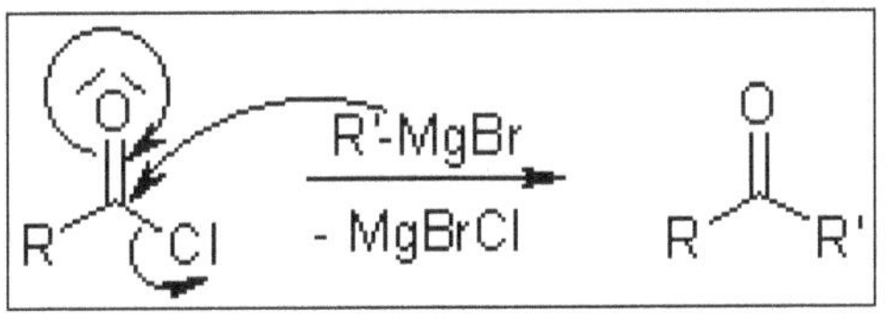

Esters are less reactive than the intermediate ketones, therefore the reaction is only suitable for synthesis of tertiary alcohols using an excess of Grignard Reagent:

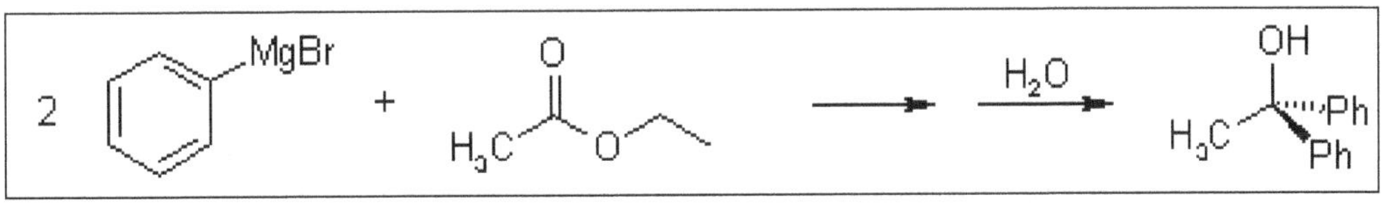

With nitriles:

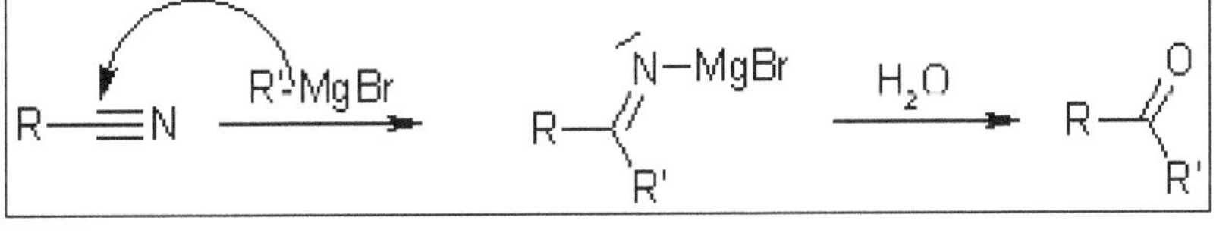

With CO_2 (by adding dry ice to the reaction mixture):

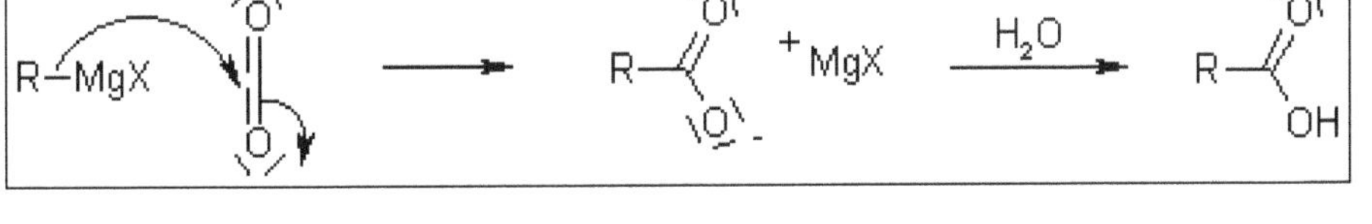

With oxiranes:

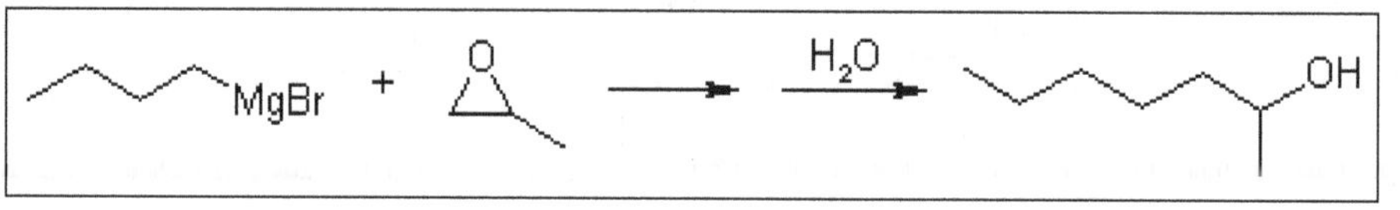

References

- Organic-reaction-mechanism-peer-reviewed-open-access-journals: omicsonline.org, Retrieved 8 May, 2019
- Acid-base-reaction: newworldencyclopedia.org, Retrieved 11 March, 2019
- Addition-reaction, chemistry: byjus.com, Retrieved 14 June, 2019
- Addition-reaction-definition-example: study.com, Retrieved 4 January, 2019
- Substitution-reaction, chemistry: byjus.com, Retrieved 17 April, 2019
- Electrophilic-substitution-reaction-mechanism, chemistry: byjus.com, Retrieved 3 January, 2019
- Introduction-to-Substitution-Reactions, Organic-Compounds, General-Chemistry: libretexts.org, Retrieved 8 July, 2019
- Elimination-reaction, chemistry: byjus.com, Retrieved 29 June, 2019
- Reimer-tiemann-reaction-mechanism: byjus.com, Retrieved 3 August, 2019
- Clemmensen-reduction: onlineorganicchemistrytutor.com, Retrieved 13 February, 2019
- Wolff-kishner-reduction: name-reaction.com, Retrieved 23 May, 2019
- Wolff-Kishner_Reduction, Reactivity_of_Aldehydes_and_Ketones, Aldehydes_and_Ketones, Organic_ Chemistry: libretexts.org, Retrieved 9 July, 2019
- The-grignard-reaction-mechanism: study.com, Retrieved 2 August, 2019
- Grignard-reaction: organic-chemistry.org, Retrieved 20 March, 2019

4

Stereochemistry: Chirality and Isomers

The sub-discipline of chemistry which studies the relative spatial arrangement of atoms in a molecule as well as their manipulation is known as stereochemistry. Chirality refers to a property of ions and molecules which are non-superimposable on their mirror images. Isomers are the molecules or ions which have the same chemical formula but different structures. The chapter closely examines these key concepts related to stereochemistry.

Stereochemistry

Stereochemistry is the study of the three-dimensional structure of molecules. The cis and trans isomers are forms of stereoisomers, differing structurally only in the location of the atoms of the molecule in three-dimensional space. Such stereoisomers can have different physical and chemical properties. For example, the cis and trans isomers of butenedioic acid show vast differences in their physical and chemical properties.

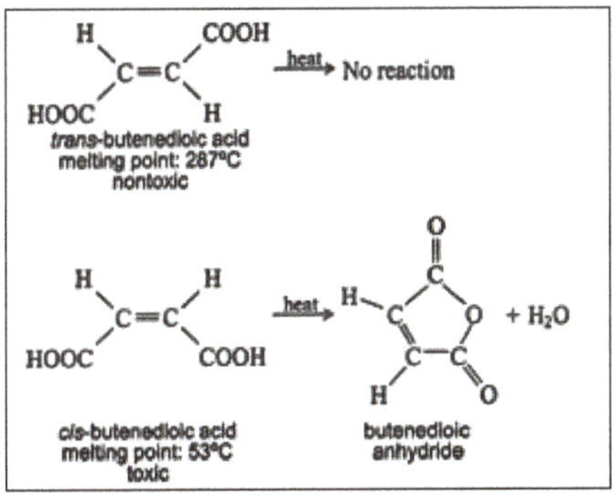

Stereochemistry is of particular interest to biochemists because the reactivity and toxicity of molecules change with their stereochemistry. Most body reactions are stereospecific, meaning that receptor sites on cells accept only molecules with specific spatial arrangements of their atoms. Other configurations of the same chemical either will not react or may be toxic to the living being.

There are other types of stereoisomers in addition to the *cis* and *trans* arrangements of atoms. To understand these other isomers, you must first understand the three-dimensional structures of the molecules.

Chirality

The geometric property of a rigid object (or spatial arrangement of points or atoms) of being non-superposable on its mirror image; such an object has no symmetry elements of the second kind (a mirror plane). If the object is superposable on its mirror image the object is described as being achiral.

A chiral center is defined as an atom in a molecule that is bonded to four different chemical species, allowing for optical isomerism. It is a stereocenter that holds a set of atoms (ligands) in space such that the structure may not be superimposed on its mirror image.

Chiral Center Examples:

The central carbon in serine is a chiral carbon. The amino group and hydrogen can rotate about the carbon.

While chiral centers in organic chemistry tend to be carbon atoms, other common atoms include phosphorus, nitrogen, and sulfur. Metal atoms may also serve as chiral centers.

A molecule is chiral if it is not superimposable on its mirror image. Most chiral molecules can be identified by their lack of a plane of symmetry or a center of symmetry. Your hand is a chiral object, as it does not have either of these types of symmetry.

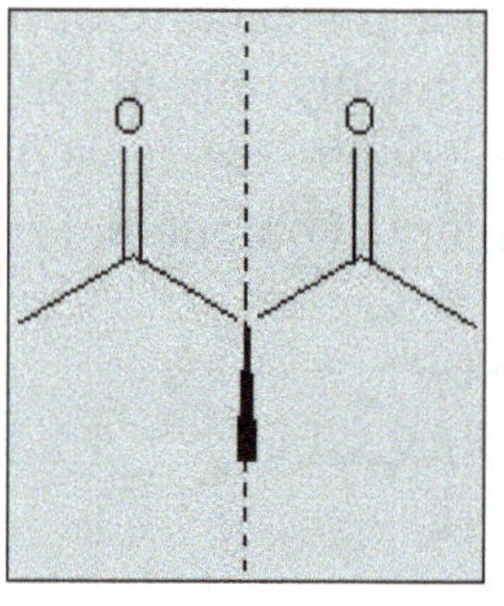

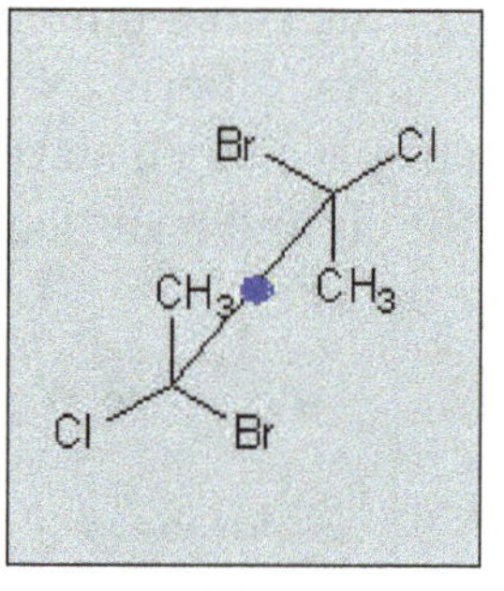

The molecule on the left has a plane of symmetry through the center carbon. This is a mirror plane; in other words, one half of the molecule is a perfect reflection of the other half of the molecule. This molecule is not chiral because of its mirror plane.

The molecule on the right has a center of symmetry, or an inversion center. An inversion center is a point in the molecule - not necessarily on an atom - through which all other atoms can be reflected 180 degrees into another, identical, atom. (In more accurate symmetry terms, an inversion through a center is equivalent to rotating groups by 180 degrees and then reflecting the groups through a plane perpendicular to the rotation axis). This type of symmetry is rare in organic molecules, and is more common in inorganic molecules. The inversion center is represented by the blue circle in the above example. The same molecule is shown three-dimensionally below. The inversion center is in the center of the middle carbon-carbon bond. This molecule is not chiral because of its inversion center.

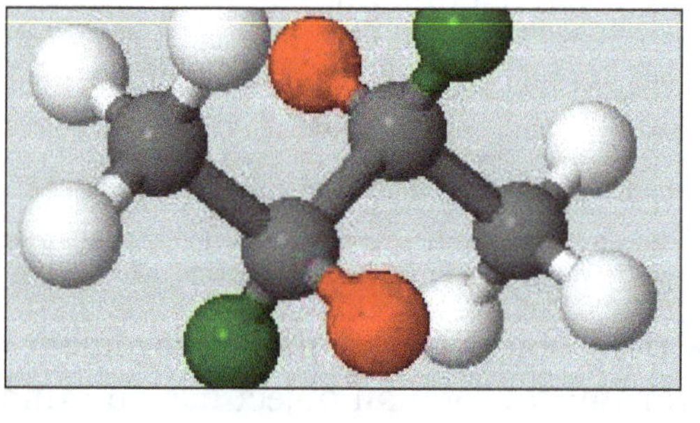

Chirality in Natural and Synthetic Materials

Much of the function of biologically active molecules depends on fit, on an exquisite lock-and-key connection between molecules that allows some biochemical activity to turn on or off. In the evolutionary process, chirality—handedness—came to be a critical part of the lock-and-key fit. The principle behind this notion is simple. A left-hand glove does not fit a right hand, and, in the same way, one member of an enantiomeric pair of molecules might fit another molecule whereas the other member would not. The specificity of biological reactions and their dependence on fit have both benefits and penalties.

As presented above, combinations of 20 possible amino acids make up proteins, and the proteins are responsible for biological function. Precise fit is critical. The lock-and-key mechanism has evolved to be extremely precise, and much of that precision is the result of the handedness of the amino acids. Precision is one of the benefits of specificity.

A classic and tragic example of the penalties of specificity is thalidomide, a compound originally marketed as a sedative in Europe in 1956. Thalidomide contains a stereogenic carbon, and therefore the compound can exist in both *R* and *S* forms. Most synthesizing procedures generate equal mixtures of enantiomers (i.e., equal amounts of the *R* and *S* forms—a racemic mixture); special care must be taken to make a pure enantiomer, and the company involved in the original promotion of thalidomide saw no reason to bear the cost of this process. The result was the marketing of a racemic mixture of the *R* and *S* forms. (*S*)-Thalidomide turned out to be a powerful teratogen: it causes all manner of external and internal abnormalities in fetuses if it is given to pregnant women in the first trimester. The *R* enantiomer is far more benign—although even it is dangerous, as the *R* form racemizes under physiological conditions and thus produces some of the dangerous *S* enantiomer.

In the example of thalidomide, the societal consequences of a bad molecular fit, as it were, have been instructional. There has been much pressure on the drug industry worldwide to do far more thorough testing and follow much better scientific procedures than was the case for thalidomide. There has also been demand for the development of the synthetic techniques necessary to produce enantiomerically pure drugs. Most syntheses in the laboratory begin with achiral materials, and the complex end products of these synthetic procedures are built up through sequences of reactions. Unless an optically active agent is introduced or a separation into enantiomers deliberately performed, any chiral end products of the synthetic sequence will be racemic mixtures. The example of thalidomide and many other similar, if not so tragic, examples have revealed the desirability of enantiospecific syntheses in drug research. Of course, it will always be necessary to test both enantiomers in case the physiological racemization discovered with thalidomide is repeated with another compound. Ironically, the notorious thalidomide is making a comeback, as it turns out to be an effective agent against several extremely difficult diseases, including leprosy and multiple myeloma. These unexpected developments are further examples of the benefits of enantiospecific synthesis.

Optical Activity

Optical activity is an effect of an optical isomer's interaction with plane-polarized light.

Isomers

An isomer is a chemical species with the same number and types of atoms as another chemical species but with distinct properties because the atoms are arranged into different chemical structures. When atoms can assume different configurations, the phenomenon is termed isomerism. There are several categories of isomers, including structural isomers, geometric isomers, optical isomers, and stereoisomers. Isomerization can occur spontaneously or not, depending on whether the bond energy of the configurations is comparable.

Isomer Examples: Pentane, 2-methylbutane, and 2,2-dimethylpropane are structural isomers of each other.

Importance of Isomerism

Isomers are especially important in nutrition and medicine because enzymes tend to work on one isomer over another. The substituted xanthines are a good example of an isomer found in food and drugs. Theobromine, caffeine, and theophylline are isomers, differing in the placement of methyl groups. Another example of isomerism occurs in phenethylamine drugs. Phentermine is a non-chiral compound that can be used as an appetite suppressant yet doesn't act as a stimulant. Rearranging the same atoms yields dextromethamphetamine, a stimulant stronger than amphetamine.

Constitutional/Structural Isomers

Isomers that differ in connectivity are called constitutional (sometimes structural) isomers. They have the same parts, but those parts are attached to each other differently. The bracelets of red and green beads are analogous to constitutional isomers. The simplest hydrocarbons—methane (CH_4), ethane (CH_3CH_3), and propane ($CH_3CH_2CH_3$)—have no constitutional isomers, as there is no other way to connect the carbons and hydrogens of these molecules consistent with the tetravalency of carbon and the univalency of hydrogen.

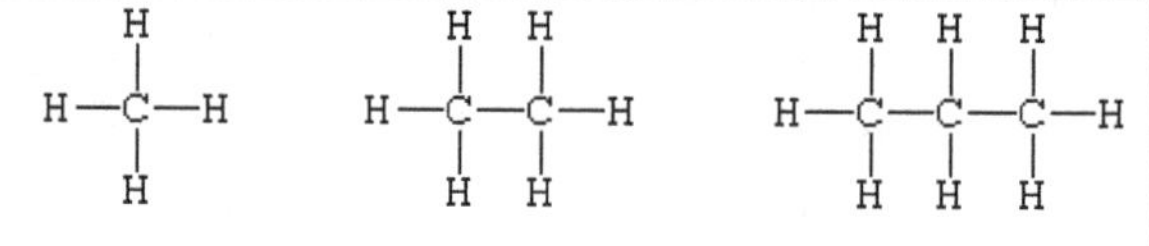

methane (CH_4) ethane (C_2H_6) propane (C_3H_8)

However, there are two different butanes, C_4H_{10}, and these two molecules, called butane and isobutane, are constitutional isomers. They are different molecules with different chemical and physical properties. Butane has its four carbon atoms bonded in a continuous chain. Isobutane has a branched structure.

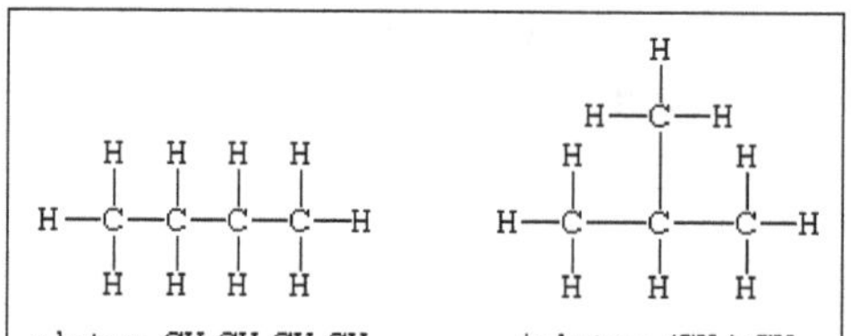

n-butane, $CH_3CH_2CH_2CH_3$ isobutane, $(CH_3)_3CH$

The number of possible constitutional isomers increases greatly with the number of available atoms. There are only two butanes, but there are three pentanes (C_5H_{12}), 18 octanes (C_8H_{18}), and no fewer than 366,319 constitutional isomers of the hydrocarbon containing 20 carbon atoms and 42 hydrogens.

Unlike conformational isomers, structural isomers differ in connectivity, as illustrated below for 1-propanol and 2-propanol. Although these two alcohols have the same molecular formula C_3H_8O, the position of the –OH group differs creating a unique compounds with differences in their physical and chemical properties.

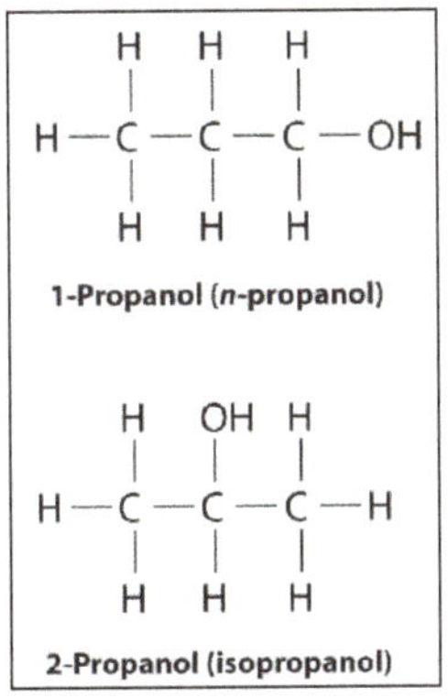

Consider, for example, the following five structures represented by the formula C_5H_{12}. In the conversion of one structural isomer to another, at least one bond must be broken and reformed at a different position in the molecule.

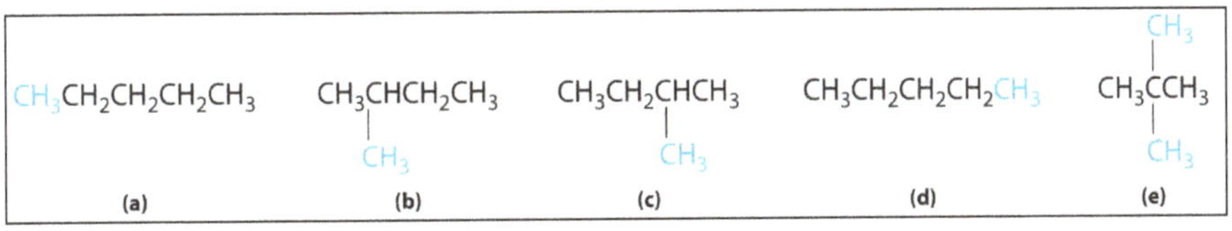

Structures above represent the same compound, n-pentane. Structures (b) and (c) represent the same compound, 2-methylbutane. No bonds need to be broken and reformed to convert between or between. The molecules are simply rotated 180° about a vertical axis. Structure is named 2,2-dimethylpropane. There are only three structural isomers possible with the chemical formula C_5H_{12}: n-pentane, 2-methylbutane, and 2,2-dimethylpropane. Structural isomers have distinct physical and chemical properties.

Types of Structural Isomerism

1. Chain, Nuclear or Skeleton Isomerism:

   This type of isomerism is due to the difference in the nature of the carbon chain (i.e. straight or branched) which forms the nucleus of the molecule, e.g. given below.

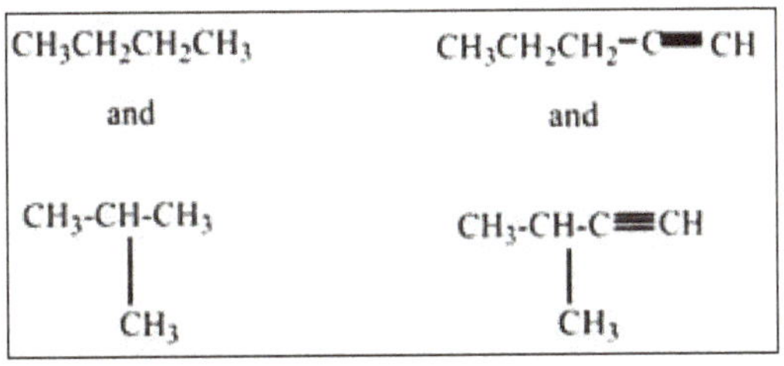

2. Position isomerism:

   It is due to the difference in the position of the side chain atom or group or an unsaturated linkage in the same carbon chain.

   Examples are:

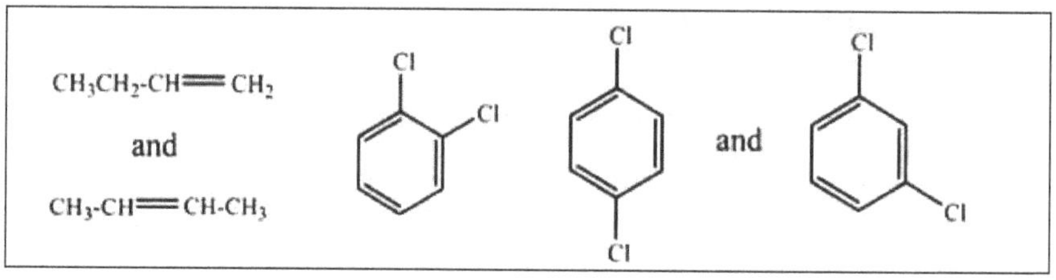

3. Functional isomerism:

   This type of isomerism is due to difference in the nature of functional group present in the isomers, e.g. given below.

C_2H_6O:	CH_3-CH_2OH Ethyl alcohol (alcoholic group)	CH_3-O-CH_3 Dimethyl ether (ether group)	
C_3H_6O:	CH_3-CH_2-CHO Propanal	CH_3-CO-CH_3 Acetone	$CH_2{=}CH$-CH_2OH Allyl alcohol

(note the different functional groups in 3 isomers)

4. Tautomerism:

   Tautomerism may be defined as the phenomenon in which a single compound exists in two readily inter-convertible structures that differ markedly in the relative position of at least one atomic nucleus, generally hydrogen. The two different structures are known as tautomers of each other. There are several types of tautomerism of which keto-enol tautomerism is the most important. In this type, one form (tautomer) exists as a ketone while the other exists as an enol. The two simplest examples are of acetone and phenol.

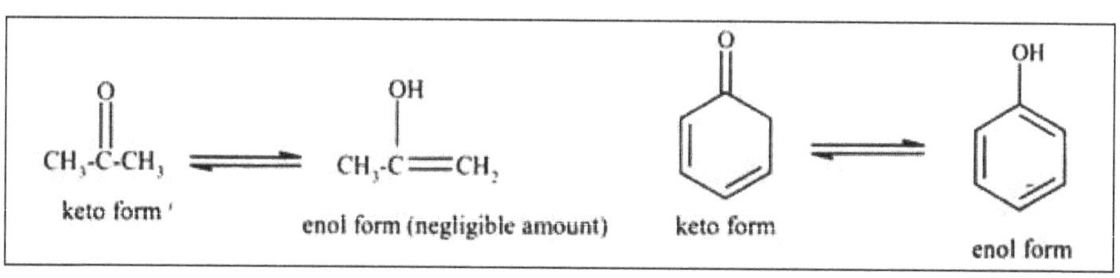

Stereoisomers

Generally defined, stereoisomers are isomers that have the same composition (that is, the same parts) but that differ in the orientation of those parts in space. There are two kinds of stereoisomers: enantiomers and diastereomers. Enantiomers are mirror images, like one's hands, and diastereomers are everything else. However, as is stated above, timescale and energy are important. In order to understand these considerations, it is helpful first to consider a special kind of stereoisomer, the conformational isomer.

Conformational Isomers

Methane (CH_4) is a molecule that is a perfect tetrahedron, and so it is commonly said that no isomerism is possible with methane. However, the carbon-hydrogen bonds of methane constantly vibrate and bend, so that on very short timescales an apparent isomerism can be said to exist. But these structures are not energy minima, and so they do not qualify as isomers.

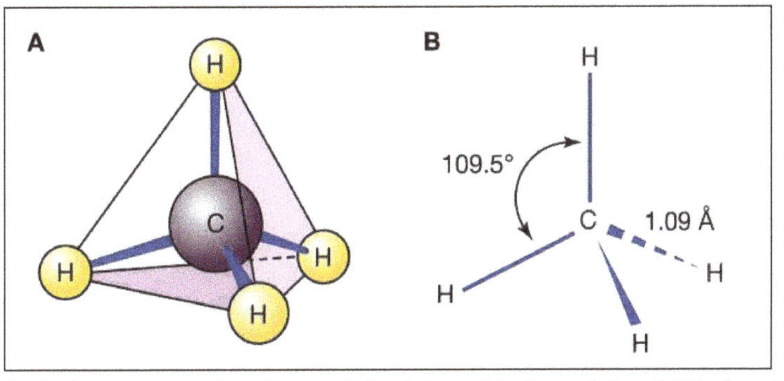

The tetrahedral geometry of methane: (A) stick-and-ball model and (B) showing bond angles and distances. (Plain bonds represent bonds in the plane of the image; wedge and dashed bonds represent those directed toward and away from the viewer, respectively).

As complexity increases, isomerism induced by rotations about bonds becomes a bigger factor. In ethane (CH_3CH_3), for example, both carbons are approximately tetrahedral. Thus, there are two limiting structures—staggered ethane, in which the carbon-hydrogen bonds are as far apart as possible, and eclipsed ethane, in which the bonds are as close as possible. These two structures are certainly not the same. Perhaps the best view in which to see the difference is a "Newman projection" (named after American chemist Melvin Newman) in which one sights down the carbon-carbon bond and focuses on the positions of the six hydrogens. In a Newman projection, the front carbon is located at the intersection of the bonds to the three attached hydrogen atoms, and the back carbon is an exploded circle, with the attached bonds emanating from the circumference of the circle.

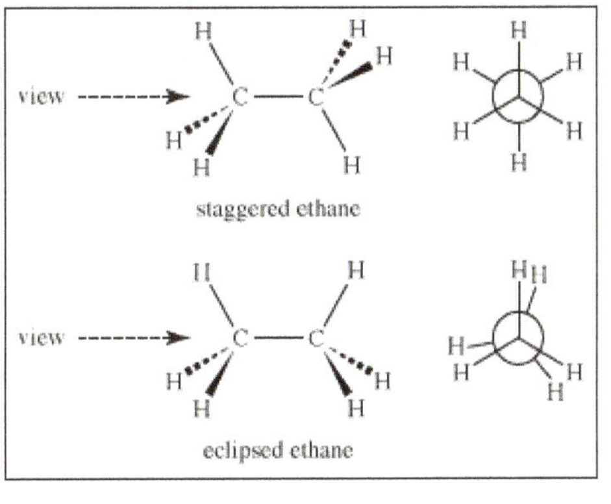

Immediately, questions of energy arise - Which of the two structures is lower in energy and therefore more stable? The staggered form is lower in energy because in the eclipsed form electrons in carbon-hydrogen bonds on the opposite side of the carbon-carbon bond repel each other. The strain that this repulsion creates increases the potential energy of the eclipsed form. The energy difference is not large, about 3 kilocalories per mole (kcal/mol).

If one plots the energy change as ethane rotates around the carbon-carbon bond, another difficulty is revealed. Like the vibrational and rotational "isomers" of methane mentioned above, eclipsed ethane (E) is not even an energy minimum; it is an energy maximum, a transition state between two staggered ethanes (S). Therefore, ethane, like methane, really has only one form.

If substitutions are made in the ethane molecule—for instance, exchanging some of the hydrogen atoms for deuterium atoms to make 1,2-dideuterioethane—isomeric staggered forms become possible. These staggered forms, called "anti" and "gauche," of 1,2-dideuterioethane are different but are interconverted through rotations around the central carbon-carbon bond and are called "conformational isomers."

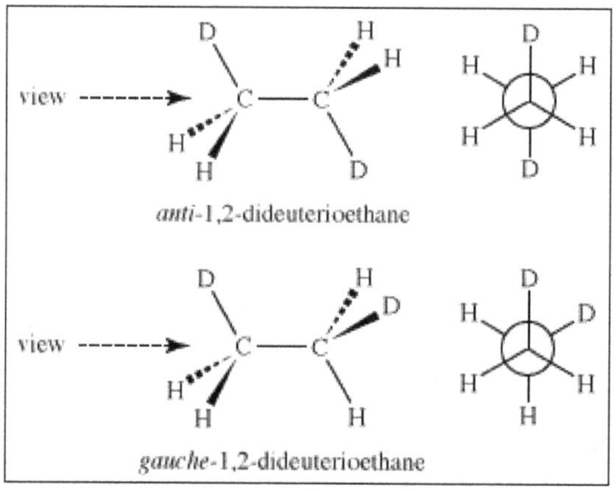

Whether these different ethanes can be separated depends only on the amount of energy necessary to convert one into the other—that is, to rotate the molecule about the carbon-carbon bond. In the case of 1,2-dideuterioethane, the energy barrier separating the conformational isomers is only 3 kcal/mol, far too low to make them separable under normal conditions.

Butane has two constitutional isomers, butane and isobutane. Isobutane has no conformational isomers, but butane is closely analogous to 1,2-dideuterioethane, in that a pair of anti and gauche conformational isomers is possible for that molecule. Because a methyl group (CH_3) is much larger than hydrogen or deuterium, the plot of energy versus rotational angle is more complex for butane than it is for ethane or 1,2-dideuterioethane.

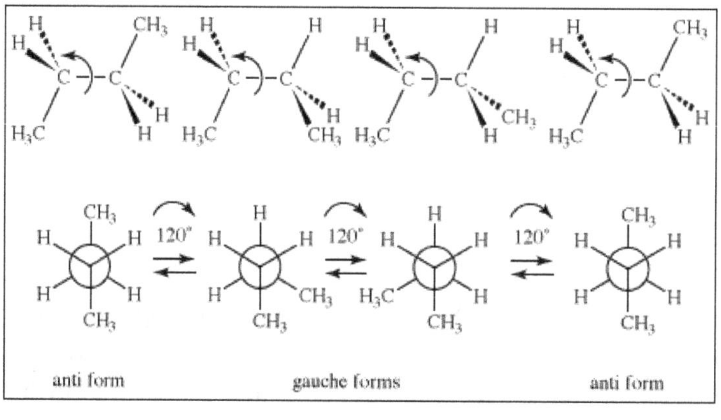

Ring compounds often have a particularly rich set of conformational isomers. By far the most interesting of the ring compounds is cyclohexane (C_6H_{12}), shown here with cyclopropane (C_3H_6).

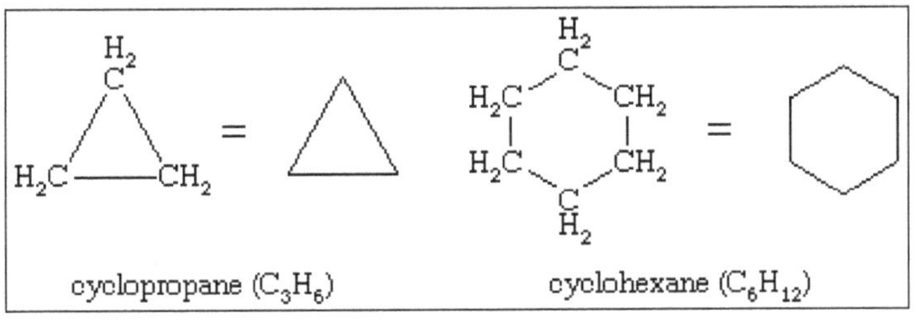

Planar cyclohexane contains 12 pairs of eclipsed carbon-hydrogen bonds and is destabilized by these eclipsing interactions, or torsional strain. There are other problems with the planar form. In a flat hexagon, the C–C–C angles must be 120°, quite far from the optimum for tetrahedral carbons (usually quoted as approximately 109.5°; in fact, the real optimum value for cyclohexane is about 112°, the C–C–C angle in propane). In any event, the planar form of cyclohexane is severely destabilized by both torsional and angle strain.

Lower-energy forms can be made as the cyclohexane ring distorts from planarity. This distortion involves no more than rotations about carbon-carbon bonds, just as occurs in ethane or any other acyclic alkane. The energy minimum for cyclohexane is the chair form. In the chair form, carbon-hydrogen bonds are nicely staggered, and the C–C–C bond angle is 111.5°, very close indeed to the optimum.

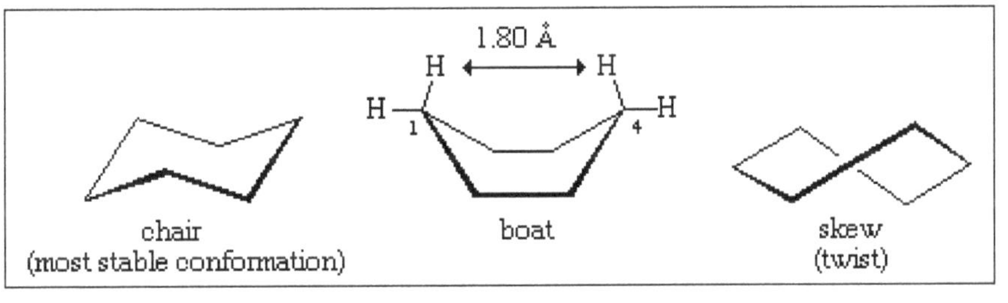

Note that there are two kinds of carbon-hydrogen bonds in chair cyclohexane. One set of six parallel carbon-hydrogen bonds is perpendicular to the surface on which the chair apparently sits (these are the axial bonds). The other set of six is roughly in the plane of the ring (equatorial bonds). All six axial hydrogens are equivalent, as are all six equatorial bonds.

Rotations about carbon-carbon bonds interconvert two equally energetic chair forms. This process is colloquially called a ring "flip."

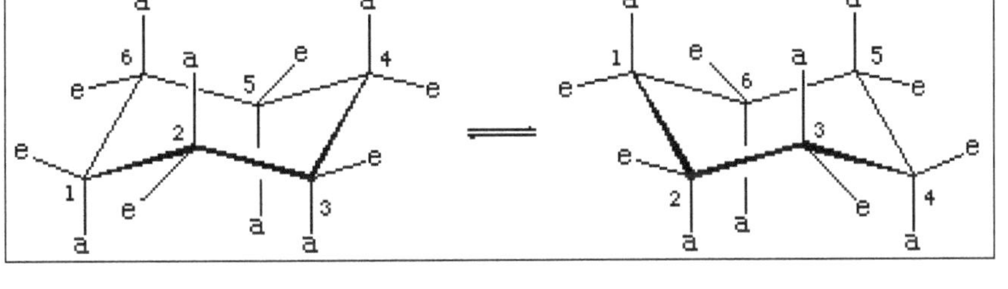

The axial hydrogens in one chair become the equatorial hydrogens in the other as the ring "flips" from one chair to the other. An implication of this change is that there is more than one monosubstituted cyclohexane. In methylcyclohexane, for example, there are two conformational isomers, one with the methyl group axial and one with the methyl group equatorial. The two interconvert through ring flipping.

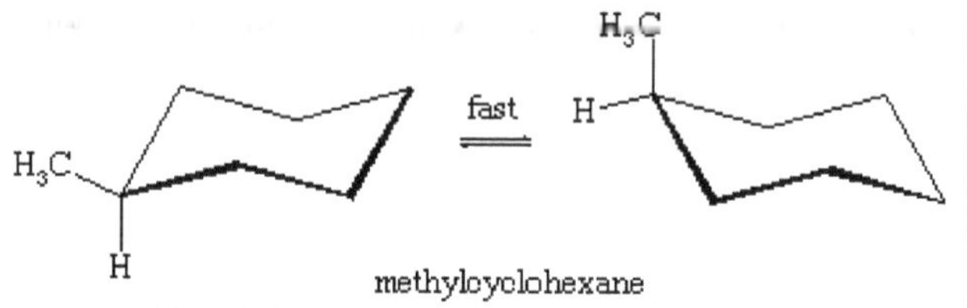

methylcyclohexane

equatorial methyl group (95 percent) | axial methyl group (5 percent)

Which isomer is more stable? For the axial methyl isomer the methyl group interacts unfavourably with nearby methylene groups. This destabilizing interaction is not present in the equatorial isomer. Axial methylcyclohexane is less stable (higher energy) than the equatorial isomer by 1.8 kcal/mol.

Cis and Trans Forms

The examples presented so far have concentrated on the simplest organic molecules, the alkanes. However, stereoisomers crop up in many of the other structural types of organic chemistry. For example, in the alkenes, two versions of 2-butene exist. They are traditionally called *cis*-2-butene and *trans*-2-butene or, in slightly more modern terms, (*Z*)- and (*E*)-2-butene. The *Z* and *E* stand for the German words for "together" (*zusammen*) and "apart" (*entgegen*). In principle, *cis*- and *trans*-2-butene are conformational isomers; in theory, they could be interconverted by a simple rotation about the central double bond. However, the practical world intrudes into principle, because this rotation would require about 66 kcal/mol, an amount of energy not available under normal conditions.

In principle, rotation interconverts these two compounds, but the rotation requires too much energy.

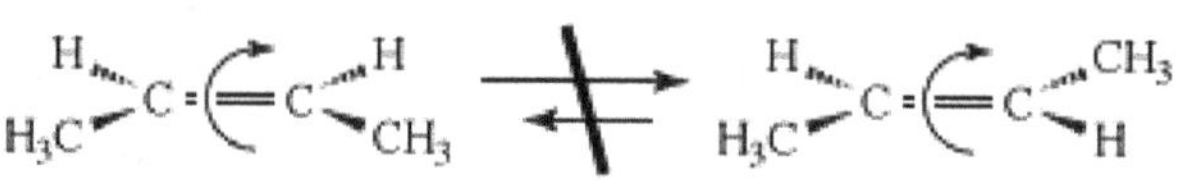

cis-2-butene | *trans*-2-butene

Other kinds of cis and trans isomers exist in ring compounds. For example, *cis*- and *trans*-1,2-dimethylcyclopropane are stereoisomers. (In the figure below, "bp" stands for "boiling point").

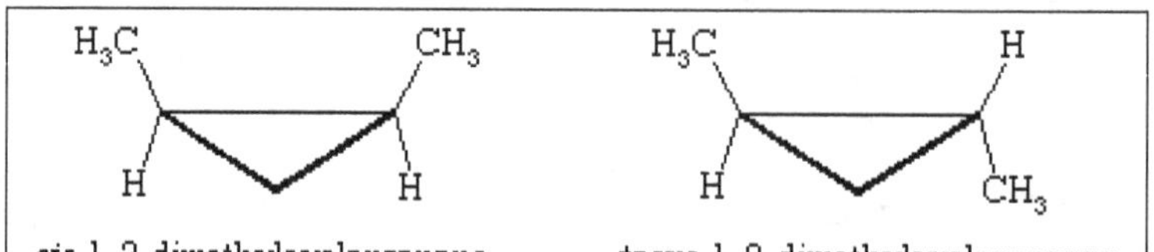

cis-1,2-dimethylcyclopropane (bp 37° C) | *trans*-1,2-dimethylcyclopropane (bp 29° C)

This time there is no imaginable rotation about bonds that can equilibrate the two isomers, so these two molecules are not conformational isomers. In addition, because three points determine a plane, the three-membered ring of cyclopropane is necessarily flat; there is no possible out-of-plane distortion.

On the other hand, cyclohexane is quite flexible, with one energy-minimum chair form ring-flipping into another through rotations around carbon-carbon bonds. Consider the possible isomers of *cis*- and *trans*-1,4-dimethylcyclohexane. If one methyl group is in the lower-energy equatorial position, then the cis compound, with both methyl groups on the same side of the ring, can be made only by placing the second methyl group in the higher-energy axial position. In constructing the trans compound, the second methyl must be placed in the equatorial position. But what happens when the ring flips? Remember that in a ring flip all axial positions become equatorial and vice versa. In the case of the *cis*-1,4-dimethylcyclohexane isomer, the equatorial-axial version flips into itself, as the axial methyl becomes equatorial and the equatorial methyl becomes axial. The two versions of *cis*-1,4-dimethylcyclohexane therefore have the same energy.

When the trans isomer flips, however, an equivalent structure is not formed, because each of the two equatorial methyl groups becomes axial. As an equatorial methyl group is more stable than an axial methyl group by 1.74 kcal/mol, the diaxial form would be less stable than the diequatorial form by about twice that amount, or 3.5 kcal/mol. In practice, this energy difference means that far less than 1 percent of the *trans*-1,4-dimethylcyclohexane present at equilibrium is in the less stable form.

Enantiomers

It is stated that one's hands are related but not the same. Exactly how are they related? Each has a thumb, little finger, and so on. Yet the hands truly are not the same, for they are not superimposable. In effect, the left and right hand are mirror images; the left hand is superimposable on the mirror image of the right hand but not on the right hand itself. Some molecules are related to their mirror images in the same manner. Such molecules are, by definition, stereoisomers, and they go by the special name of enantiomers.

The phenomenon of handedness, or "chirality," is perhaps the most important phenomenon related to isomerism. Many objects in the macroscopic world are chiral. A scissors and a screw are familiar chiral objects; they are not superimposable on their mirror images. But related objects, a simple knife or a nail, for example, are superimposable on their mirror images and thus are not chiral. The simple test for chirality is the same for objects as it is for molecules: Is the object (or molecule) superimposable on its mirror image? If it is not, it is chiral. If it is, the object is achiral (not chiral).

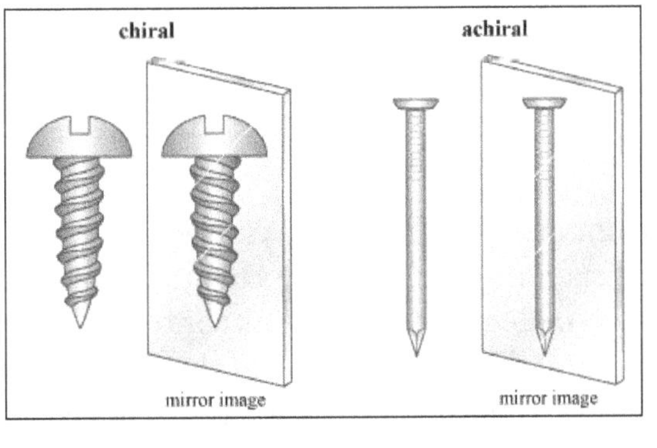

If each hydrogen atom in a molecule of methane were replaced with a different atom, one possible result would be bromochlorofluoroiodomethane (CBrClFI). The mirror images of this molecule are not superimposable. There are definitely two enantiomers of this molecule.

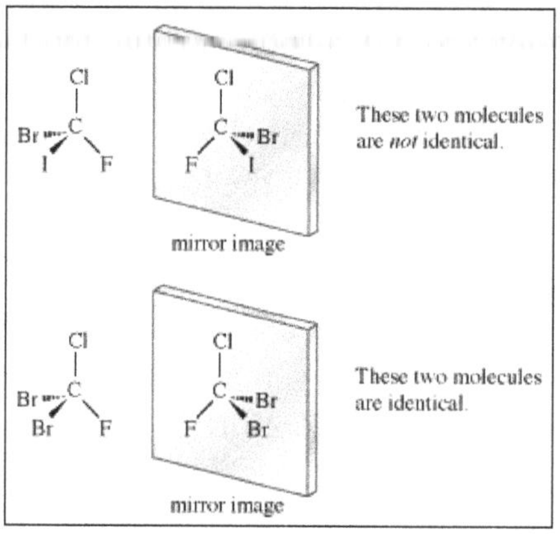

The molecule, with four different atoms (fluorine, chlorine, bromine, and iodine) attached to what is called a stereogenic carbon, is identified as chiral. Several questions and problems now appear. How are the two enantiomers different physically and chemically? On a more mundane level, how can one specify in words one of the two enantiomers? How is one to differentiate the right-handed molecule of bromochlorofluoroiodomethane from the left-handed version, for example? A rather complex protocol has been devised, and it is worth giving a simplified version of it here. In the so-called Cahn-Ingold-Prelog (CIP) protocol (named after British chemists Robert Cahn and Sir Christopher Ingold and Swiss chemist Vladimir Prelog), one first assigns priorities to the four atoms attached to the stereogenic atom, in this case the carbon at the centre of the tetrahedron. The atom of lowest atomic number is given the lowest priority, 4. In this case that atom is the fluorine (atomic number 9). The atom with the highest atomic number, iodine (atomic number 53), gets the highest priority, 1. Chlorine (atomic number 17) is priority 3, and bromine (atomic number 35) is 2. In the second step of the protocol, one sights down the bond from carbon to the lowest priority. Finally, one connects atoms 1 to 2 to 3 with an arrow. If that arrow is clockwise, the molecule is called an *R* enantiomer. If the arrow is counterclockwise, the molecule is called an *S* enantiomer. The CIP protocol is simple in the example chosen, but very often more-detailed rules must be applied to assign *R* and *S*.

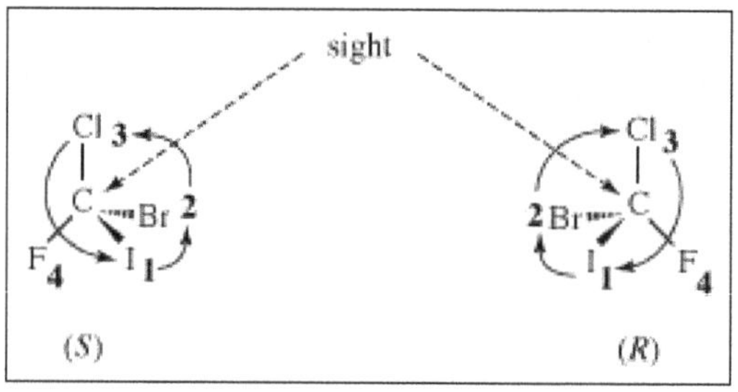

In summary, a molecule with one stereogenic carbon can be either *R* or *S*—in a sense, "left-handed" or "right-handed." One set of such molecules is of profound biological importance: the L-amino acids. The business of the body—biological function—is controlled by proteins, which are polymers of only 20 possible amino acids. All the amino acids but the achiral glycine contain a single stereogenic carbon, and all of them but glycine are left-handed. It is not clear why this sense was selected through evolution; perhaps that selection followed from an initial accident selecting for left-handed amino acids. Extraterrestrial life (should there be any) may well be either left- or right-handed.

All physical attributes of enantiomers are identical except for one rather arcane property: the direction of rotation of the plane of plane-polarized light. If one enantiomer rotates the plane in one direction (say, clockwise) as one view the beam, the other enantiomer will rotate the plane by the same amount in the other direction (in this case, counterclockwise). For example, the rotations of standard solutions of (*R*)- and (*S*)-2-aminobutane are −7.4 and +7.4 degrees, respectively.

The chemical properties of enantiomers are also identical, as long as the other reacting molecule is achiral. Chemical reactions of enantiomers are analogous to a hand grasping a ball. If the ball is featureless, one's right and left hands (enantiomers) have exactly the same interactions with the ball. However, if the ball has the word "Label" written on it and is thus a chiral object, that labeled ball will be optically active; it will mimic a single enantiomer. The left and right hands will interact differently with the labeled ball. The little finger of the right hand will approach the capital "L" of "Label" while the thumb will approach the lowercase "l." For the left hand, the interactions will be just the opposite: the little finger will approach the lowercase "l" and the thumb the capital "L."

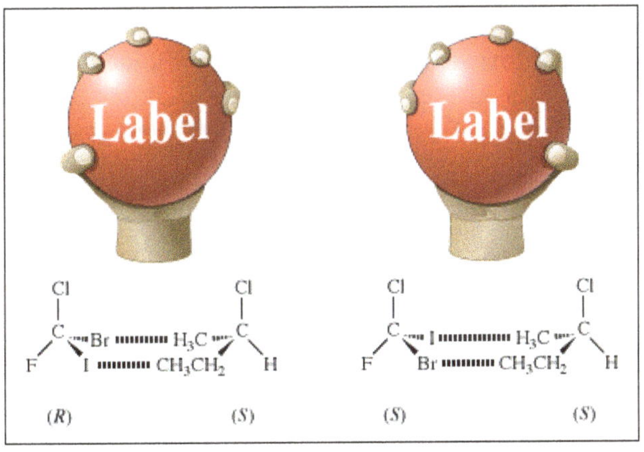

The figure shows these interactions, as well as a molecular counterpart in which a pair of enantiomers, (*R*)- and (*S*)-bromochlorofluoroiodomethane, interacts with a single enantiomer, (*S*)-2-chlorobutane. In the *R* enantiomer, the bromine atom approaches the methyl group ($-CH_3$), and the iodine atom approaches the ethyl group ($-CH_2CH_3$). In the *S* enantiomer, the bromine atom approaches the ethyl group, and the iodine atom approaches the methyl group. The figure shows only one of the many possible interactions, every one of which is different.

Stereoisomers of more Complex Molecules

An atom is stereogenic if switching any two atoms or groups of atoms that are bound to it results in a pair of stereoisomers. Very often the situation is more complex; indeed, there can be several stereogenic atoms in a molecule. A molecule with only one stereogenic atom has only two stereoisomers—the *R* and *S* enantiomers. If there are two stereogenic atoms in a molecule, both can be either *R* or *S*. Thus, there are four possibilities: *RR*, *SS*, *RS*, and *SR*. Three stereogenic atoms would lead to eight possibilities: *RRR*, *RRS*, *RSR*, *SRR*, *SSR*, *SRS*, *RSS*, and *SSS*. The formula for finding the maximum number of stereoisomers X is $X = 2^n$, where n is the number of stereogenic atoms in the molecule.

The formula $X = 2^n$ reliably gives the maximum number of stereoisomers, but in situations of high symmetry it fails to give the real number. For example, it fails for 2,3-dichlorobutane [$H_2Cl_2(CH_3)_2$]. One pair of enantiomers, *SS* and *RR*, does appear. But the other combination gives an identical "pair" of *SR* compounds. This happens because 2,3-dichlorobutane contains an internal plane of symmetry. The result is fewer than the maximum number of stereoisomers predicted by the formula. Three stereoisomers are possible: one pair of enantiomers (A and B) and an achiral molecule C, called a "meso compound." A meso compound is an achiral molecule that nonetheless contains a stereogenic atom.

In order to find molecules that are enantiomers, one must draw the mirror image of the original and see if they are superimposable. That is the only absolutely safe way to do it. It might be suggested that there is something special about a molecule containing four different groups attached to one carbon. The question now is whether the presence of such an atom (usually carbon) is either sufficient or necessary for the molecule to be chiral. The answer is no in each case. Although looking for such carbons is a good way to start a search for enantiomers, there is no way to avoid the ultimate necessity of writing out the mirror image and checking for superimposability. To test the question of sufficiency, for example, look at the meso compound C of 2,3-dimethylbutane. It certainly does contain a carbon attached to four different groups. The indicated carbon C_2 is attached to hydrogen, a methyl group, a chlorine, and the rest of the molecule. Yet C is achiral.

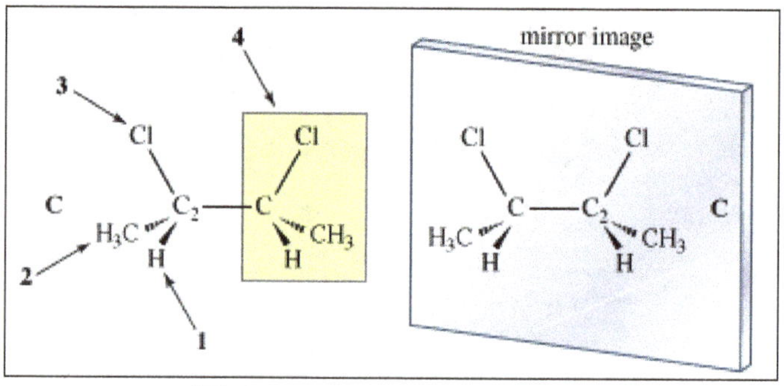

Four different groups attached to one carbon
but still superimposable on its mirror image.

There are many compounds whose molecular architecture makes them chiral but that do not contain an atom attached to four different groups. One classic example is hexahelicene, a molecule composed of six benzene rings connected to each other. The molecule coils in the form of a spiral so that the atoms of the last ring do not impinge on the atoms of the first ring. The result is a left- or right-handed screw form, and the molecule is chiral.

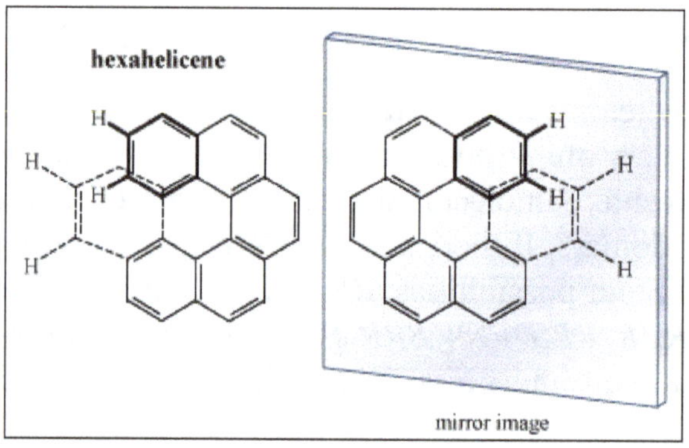

Diastereomers

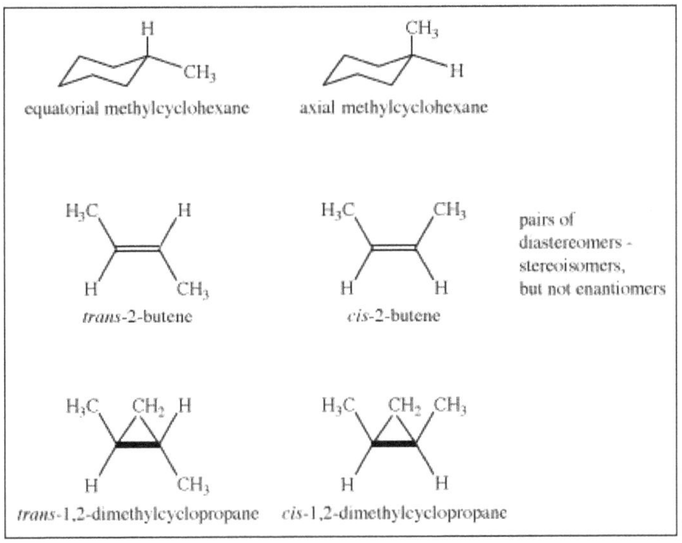

Cyclohexane is achiral, as are both axial and equatorial methylcyclohexane. The two methylcyclohexanes (axial and equatorial methyl group) are stereoisomers, but they are not enantiomers. Such isomers—stereoisomers that are not mirror images—are called diastereomers. The molecules *cis-* and *trans*-2-butene are diastereomers, as are *cis-* and *trans*-1,2-dimethylcyclopropane. However, in dimethylcyclopropane, the cis compound is achiral, but the trans compound exists as a pair of enantiomers. Therefore, there are three stereoisomers of 1,2-dimethylcyclopropane.

References

- Stereochemistry-defined, stereochemistry, organic-chemistry, chemistry: cliffsnotes.com, Retrieved 25 February, 2019
- Definition-of-chiral-center: thoughtco.com, Retrieved 23 June, 2019
- Chirality-in-natural-and-synthetic-materials, isomerism, science: britannica.com, Retrieved 15 March, 2019
- Definition-of-isomer: thoughtco.com, Retrieved 7 May, 2019
- Isomerism, science: britannica.com, Retrieved 17 July, 2019
- Types-of-isomerism-chemical-compounds-chemistry, isomerism, biochemistry: biologydiscussion.com, Retrieved 13 April, 2019

5

Applications of Organic Chemistry

Organic chemistry is used in a variety of different fields such as medicine and food industry. Some of the areas within food industry where organic chemistry is particularly used are coloring chemicals, food emulsifiers, flavor enhancing agents and moisture preservers. The diverse applications of organic chemistry in these fields have been thoroughly discussed in this chapter.

There are numerous examples of organic chemistry in the world.

Polymers

Polymers consist of long chains and branches of molecules. Common polymers you encounter every day are organic molecules. Examples include nylon, acrylic, PVC, polycarbonate, cellulose, and polyethylene.

Petrochemicals

Petrochemicals are chemicals derived from crude oil or petroleum. Fractional distillation separates the raw material into organic compounds according to their different boiling points. You encounter products made from petrochemicals every day. Examples include gasoline, plastics, detergents, dyes, food additives, natural gas, and medicines.

Soaps and Detergents

Although both are used for cleaning, soap and detergent are two different examples of organic chemistry. Soap is made by the saponification reaction, which reacts a hydroxide with an organic molecule (e.g., an animal fat) to produce glycerol and crude soap. While soap is an emulsifier, detergents tackle oily, greasy (organic) soiling mainly because they are surfactants.

Perfume

Whether a fragrance comes from a flower or a lab, the molecules you smell and enjoy are an example of organic chemistry.

Cosmetics

The cosmetic industry is a lucrative sector of organic chemistry. Chemists examine changes in the skin in response to metabolic and environmental factors, formulate products to address skin problems and enhance beauty, and analyze how cosmetics interact with the skin and other products.

Examples of Common Organic Chemicals:

- Shampoo
- Enzymes

- Gasoline
- Perfume
- Lotion
- Drugs
- Food and food additives
- Plastics
- Paper
- Insect repellent
- Synthetic fabrics (nylon, polyester, rayon)
- Paint
- Moth balls (naphthalene)
- Nail polish remover
- Wood
- Coal
- Natural gas
- Solvents
- Fertilizers
- Vitamins
- Dyes
- Soap
- Candles
- Asphalt

Organic Chemistry in Medicine

Paracetamol

Paracetamol is List of Organic Chemicals, the common name for an organic compound, N-acetyl-para-aminophenol. It is used as analgesic drug. Analgesic is a term used for drugs which have function to relieve pain. You may also refer analgesic drugs as painkillers.

Not only works as a painkiller, Paracetamol is also used to reduce fever, make it an analgesic-antipyretic drug. Nowadays, Paracetamol is widely used as medicine used to reduce headache. Indeed, that's the uses of organic chemistry in medicine.

Aspirin

Aspirin or acetyl salicylic acid, just like Paracetamol, is an analgesic-antipyretic drug. Given in higher doses, aspirin can be used as anti-inflammatory to treat arthritis. Aspirin can also be used for treating angina. Its function is to prevent the blockage of blood vessels within heart due to blood clotting.

Aspirin has several side effects. It can cause an upset stomach whether usual pain or even worse, the stomach bleeding. To prevent these side effects, you better consume them after eating your food. Other side effects caused by high doses of Aspirin include ringing in the ears, blurred vision and even dead.

Antalgin

Antalgin or Methampyron is a common name for sodium-2,3-dimethyl-1-phenyl-5-methyl-4-pyrazolon aminoetanosulfonat. Antalgin is also part of analgesic-antipyretic drugs used for reducing pain. Antalgin, along with Paracetamol and Aspirin, is the three most well-known of analgesic-antipyretic drugs. Other organic compounds which are used for some purpose are Acetosal and Phenobarbital.

Fluorouracil (5-FU)

You may have heard much about cancer. Do you know that organic compound is also used for treating cancer? The organic compound we talked about is 5-Fluorouracil. 5-Fluorouracil is injected to cancer patient's vein, including stomach cancer, pancreatic cancer, breast cancer, cervical cancer and others.

Side effects that you should aware are inflammation of the mouth and skin, loss of appetite, hair loss and low number of blood cells. These are common happened to people with cancer. For use in skin as a cream, it may cause irritation.

Sulfadiazine

Sulfadiazine belongs to organic compound which is known as aminobenzenesulfonamides. Half of these compound is containing benzenesulfonamide with an amine group attached to its benzene ring. This is an anti-bacterial drug used for the treatment of rheumatic fever and meningococcal meningitis.

An anti-bacterial drug is a part of antibiotic drugs. Antibiotic drugs are produced by organisms, mainly fungus and soil bacteria for treating diseases caused by bacteria and viruses. Other organic compounds used as antibiotic drugs are Tetracycline and Ampicillin. Tetracycline is used to treat pneumonia, whereas Ampicillin is used to prevent and treat meningitis.

Isoprenaline

Isoprenaline is classified as a part of sympathomimetic drugs. You may unusual with this term. Sympathomimetic drugs are stimulant organic compound that can imitate the works done by sympathetic nervous system.

Isoprenaline, also known as isoproterenol, is the isopropylamine analog of epinephrine (adrenaline) inside human body. Based on this spesification, Isoprenaline is used for the treatment of bradycardia (slow heart rate) and heart block. Sometimes, it is also used for treating asthma by using metered aerosol or nebulizing devices.

Other useful of this organic compound is to treat Adams-Stokes syndrome's patients, as cardiogenic shock, cardiac arrest and for treating bronchospasm occurred during anesthesia. Although has several uses, the use of Isoprenaline should be followed by some cautions. It should not be used in people with tachycardia or heart block caused by digitalis poisoning or with angina.

Prednisolone

11,17-Dihydroxy-17-(2-hydroxyacetyl)-10,13-dimethyl-6,7,8,9,10,11,12,13,14,15,16,17-dodecahydrocyclopenta[a] phenanthren-3-one, easily called as Prednisolone, is a corticosteroid medicine. Before talking about the uses, we must first know about corticostreroids. It is a class of steroid hormones, produced in the adrenal cortex of vertebrates.

Prednisolone is used to treat allergies, autoimmune disorders and even cancer. As allergies treatment, Prednisolone is useful for a wide range treatment, ranging from seasonal allergies to drug allergies. The compound also brings side effects, such as nausea and feeling tired. For a long-term use, the common side effects are including bone loss and weakness.

Chloroform

Chloroform, an organic compound with chemical formula $CHCl_3$. Related to medicine, it is used as anesthetic. Anesthetic is a drug classification which is used to prevent pain during surgery.

Anesthetic works biochemically causing cell depression, mainly to the neural network. Other organic compounds that are commonly used as anesthetic are ether and ethyl chloride. Among those three, chloroform is the strongest anesthetic, which is five times stronger than Ether. Meanwhile, ethyl chloride is the least in form as anesthetic. It is usually used for local anesthetic and small surgery.

Turmeric

Turmeric has been used as medicine for treating variety of internal disorders, including indigestion, throat infections, common colds up to cleanse wounds. These uses of Turmeric has been known for long time ago. It is a traditional medicine used by people.

Betel Leaf

Another traditional and also herbal medicine that is widely used by people is Betel Leaf. Betel Leaf contains high antiseptic substances used to overcome body and mouth odor, mouth sores and nosebleeds. Nowadays, there are so many products which are sold in markets that use Betel Leaf as extract for cleansing vaginal discharge in women. This useful of Betel Leaf has already been clinically tested.

Antiseptic substances are used to kill germ. The common substance used as antiseptic is phenol. Antiseptic substances in Betel Leaf, referred to nature antiseptic, are five times more effective than the ordinary phenol. Other uses are including herbal medicine for cough, bronchitis, burn, itchy and read eye, ulceration, stopping bleeding gums, breastfeeding to reduce excess and even pimple.

Organic Chemistry in Food Industry

Nowadays, food industry does not only rely on the non processed stuffs but also use chemical either to enhance the flavor or appearance. There are some importance of organic chemistry in food industry.

Preservative Agent

In modern chemistry, artificial preservative is more common since it is cheaper. However, preservative from organic compound is still widely used and people consider it safer. The common natural preservative for example is salt and frozen hydrogen oxide.

Many types of food especially in traditional serving use salt as preservative to keep the food edibility longer. Also, the method of using ice as preservative is still the main mechanism by freezing the items in cold temperature. Thus, organic chemistry from naturally occurring chemical is important in food industry.

Coloring Chemical

The ancient method of enhancing the appearance of the food to make it more appealing and appetizing used color agents from natural source. As science evolved, even some artificial color additives are still derivation of organic compounds.

In facts, many health association of various countries suggest food industry to use organic color additive as it is safer for human health. Some of the most famous color additive for food are carotenoids for yellow, orange, and deep red derived from carrot, potato, and pumpkin and chlorophyll from plants' leaves for green color.

Crop Protection

Crops and other plantation products need high maintenance from unwanted disturbance like pest and weeds. Therefore, in order to get the stock stable and the quality good, farmers need crop protection. Chemical generated from organic compound is the best pesticide as it can blend better with the environment without causing multitude damage to ecosystem. Even though artifical pesticide is more popular, the one from organic chemistry is more recommendable for the long sustain of food industry.

Crop Fertilizer

Another big contribution of organic chemistry in food industry is as crop fertilizer. Many foods are result of processed natural source such as crops, vegetables, fruits, seeds, and many more. In order to increase the production, farmer needs fertilizer.

Synthetic fertilizer sells better but organic fertilizer is safer as well as less costly. Basically any living being that contains nutrient needed such as nitrogen, potassium, ammonium nitrate and the like can be fertilizer. Farmers can take it from animal feses, seaweeds, fish extracts and many more.

Food Emulsifier

Food emulsifier is important in the making of food to combine water and oil element of the food materials as the two naturally will not mix together. Emulsifier helps to make a more various food that combines ingredients from both water and oil based.

Organic emulsifier derivate from organic chemistry such as milk, phospholipids chemicals, sorbitan, sodium, and eggs. In facts, many emulsifier are from natural sources as they blend with the food better with almost no unpleasant after taste.

Flavor Enhancing Agents

Popular flavor enhancer monosodium glutamate comes from organic compound when sugar elements are distilled and processed to enhance the papillae on human's tongue. Spices that generally favored for its rich taste also come from natural source. compared to the synthetic ones, organic flavor enhancers are more various as human has used it ever since they know cooking.

Therefore, the role of organic chemistry in developing flavor enhancer is very important in food industry as people like to have a tastier meal.

Moisture Preserver

While organic preservative mostly works by drain out the liquid of the food substance, moisture preserver helps to keep the moist within food or food ingredients. Most common food source that needs moisture preserver is hay and leaves.

And popular moisture preserver such as propionic acid often sells for animals feeding. Organic chemistry has wider branch and possibility as it has open source. Therefore, moisture preserver in nature is easier to look for than by creating it in laboratory.

Food Additives

A food additive is anything that affects food (directly or indirectly) or is a component of food. Legally, a food additive is anything added to food, or used in food preparation, that is not on the 'Generally Recognized as Safe' (GRAS) list. The GRAS list includes products such as flour, sugar, and salt - any ingredient that has been used for a long time and has shown no adverse effects; so food additives are ingredients that need government approval before they can be added to food.

Some foods, such as sugar, are included on the GRAS list because they have been used for years with no adverse effects seen. When it comes to food additives, there is a lot of controversy over how safe they are and if they should be used.

Categories of Food Additives

Acids

As shown in Table, most of the food acids are tracked down to the fruits such as grapes, lemon, orange, berries, plums, etc. While the human metabolic system readily adjusts to most of the natural food acids, there are some acids that the human body does not naturally recognize and are unhealthy thus leading to irritation and inflammation.

Table: List of Food Acids.

Acid	Source
Citric Acid	Citrus fruits – Lemon, Orange
Malic Acid	Apple
Tartaric Acid	Grapes, Pineapples
Acetic Acid	Vinegar
Oxalic Acid	Tea, Pepper
Tannic Acid	Tea
Caffeotannic Acid	Coffee
Benzoic Acid	Cranberries, Plums
Butyric Acid	Decomposition of Butter
Lactic Acid	Butter Digestion

Citric Acid (IUPAC Name)

2-hydroxypropane-1, 2, 3-tricarboxylic acid) - It is a weak organic acid with the chemical formula $C_6H_8O_7$. Its position is well established in the food industry as a natural preservative and additionally it gives a characteristic acidic or sour taste to food and drinks. When heated above 175 °C, it undergoes decarboxylation i.e. loss of carbon dioxide. The acid strength of citric acid is somewhat higher than typical carboxylic acids. Applications: (a) Citric acid is the acid of choice in the beverage industry and is used widely in carbonated beverages for its flavoring and buffering (pH maintaining) properties while its solubility properties make it a supreme additive for syrup concentrates; (b) It also increases the efficacy of anti-microbial (bacterial, fungal, etc.) preservatives. Citric acid (sometimes used as its citrate salt) is also used in non-carbonated beverages such as juices, thirst-quenchers, etc. (c) It is used in dry powder beverages, as also inartificially sweetened beverages. Citric acid helps add bulk and mouth-feel typically attributed to sucrose. In a recent paper the addition of citric acid showed reduced levels of the carcinogen acrylamide in cooked food.

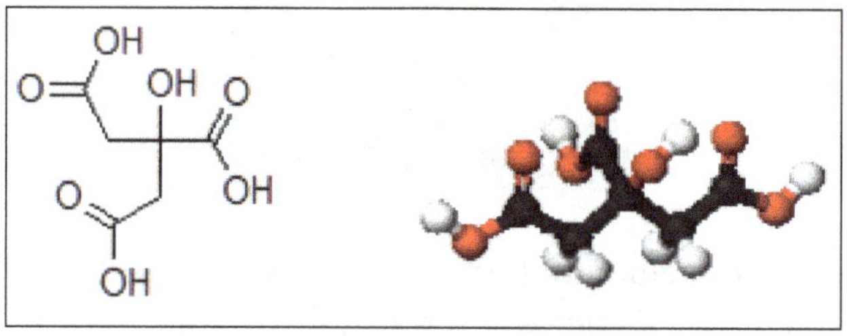

Figure: Structure of citric acid.

Tartaric acid (IUPAC Name)

2, 3-dihydroxybutanedioic acid) - Tartaric acid is a white crystalline dicarboxylic aldaric acid with the formula $C_4H_6O_6$. It occurs naturally in many plants, e.g. grapes, bananas, and tamarinds. It functions as a leavening agent when added together with baking soda; is also routinely added to wine as the principal acidulant; is used as an antioxidant which makes it a more healthy alternative acid. Occasionally, tartaric acid is replaced by its salt i.e. tartarate. Chemically, it is a dihydroxy derivative of succinic acid which is itself used as a buffering agent in foods.

Figure: Structure of tartaric acid.

Malic Acid (IUPAC Name)

2-hydroxybutanedioic acid)-Malic acid is an organic compound with the formula HO_2CCH_2-$CHOHCO_2H$ ($C_4H_6O_5$) and exists in its two stereoisomeric forms (L- and D-enantiomers), though only the L-isomer is found in nature. The salts of malic acid are known as malates. Applications: (a) low calorie beverages -its biggest advantage is that as compared to citric acid, the amount of malic acid required to be added to achieve the same level of sourness, is significantly lower. Malic acid has an extended sourness i.e. its sourness stays longer and hence results in a more balanced taste. (b) Hard candy-It melts sooner than other food acids and hence can be assimilated easily into the molten hard candy without adding any extra water thus leading to increased shelf-life. (c) Chewing gum-For good chewing gum, it is important that it lead to profuse secretion of saliva and to achieve it, a combination of saccharin and food acid is often used. Blends of malic acid with other food acids having different miscibilities results in a progressive release of the acid creating sustained juiciness and flavor during chewing. (d) Desserts - it enhances fruit flavor in sherbets and water ices. In gelled desserts, addition of controlled amounts of malic acid leads to enhanced fruit flavor and pH control.

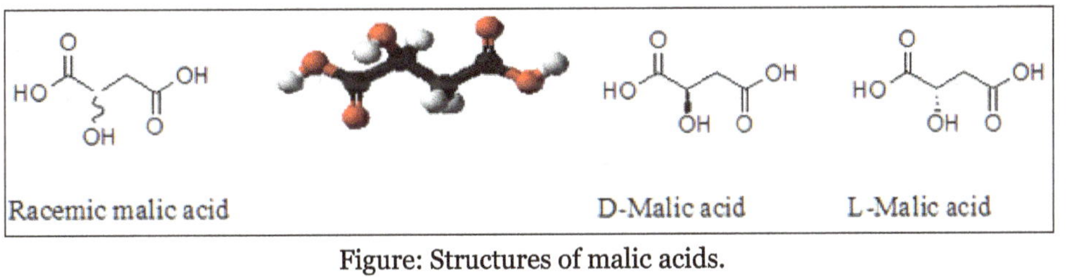

Figure: Structures of malic acids.

Fumaric Acid (IUPAC Name)

(E)-butenedioic acid) - Fumaric acid is the chemical compound with the formula $HO_2CCH{=}CH$-CO_2H. Among all the food acids we have discussed up to now, this is the only acid having a double bond in its structure and it is therefore classified as an unsaturated acid. This white crystalline acid is one of the two isomeric unsaturated dicarboxylic acids, the other being maleic acid (the cis-isomer). Fumaric acid has a natural fruity taste and its salts and esters are called fumarates. Incidentally, dimethyl fumarate reduces disability progression in multiple sclerosis.

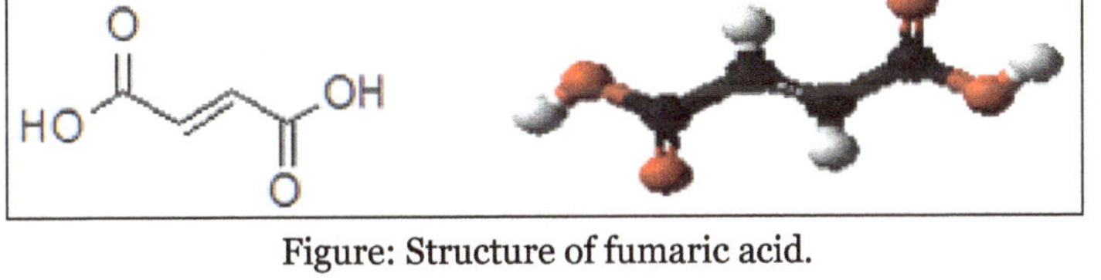

Figure: Structure of fumaric acid.

Applications

As a food additive, this acid is mainly used as an acidity regulator. Fumaric acid is a food acidulent used since 1946. It is usually a good alternative acid for tartaric acid and sometimes of citric acid. When used as a replacement for citric acid, it is added at a rate of 1.36 g of citric acid to every 0.91 grams of fumaric acid. It is also the acid of choice as a coagulant in stovetop pudding mixes.

Lactic Acid (IUPAC Name)

(2-hydroxypropanoic acid): Lactic acid, (obtained from milk and hence termed as "milk acid"), is an important chemical compound that plays a crucial role in various biochemical processes. Lactic acid is a a-hydroxy carboxylic acid with the chemical formula $C_3H_6O_3$. It exists in two enantiomeric forms: L (+)-lactic acid and D (-)-lactic acid, the former being the enantiomer that is primarily involved in human metabolism. In early 1960's, the presence of D-lactic acid in infant milk formulas was found to lead to infant acidosis. Applications - Being widely prevalent in natural foods, it is extant in many foodstuffs. Natural fermentation of mainly milk-based products e.g. cheese, yogurt, soy sauce, meat products, etc yields lactic acid and hence it is widely used in food applications such as confectionery, bakery products, meat products, beverages, dairy products, salads, dressings, ready to-eat meals, etc.

Lactic acid when used as a food additive mainly serves as a pH regulator or a preservative. Sometimes, it is also used as a flavoring agent. Among its many uses are: (a) Meat, Poultry and Fish: in the form of its salts sodium or potassium lactate, it is employed to extend shelf life, control disease-causing bacteria (thus improving the safety and nutritional value of food), enhance (and protect) meat flavor, and improve water binding/retention capacity. (b) Pickled Vegetables: here again the natural preservative action of lactic acid is employed in preventing the damage of olives, gherkins, pearl onions (pearl onions closely resemble leek and are sweeter than common onions), and other vegetables preserved in brine. (c) Salads and Dressings: addition of lactic acid provides a relatively milder flavor to food products while maintaining natural microbial stability and safety. (d) Baked Goods: Sourdough is a type of bread made by prolonged fermentation of dough with lactic acid where the latter is used directly for acidification in the production process. (e) Savory Flavors: Lactic acid does not just enhance the usual flavors, but it also enhances a broad range of savory flavors (also called umami flavors).

Figure: Structure of lactic acid.

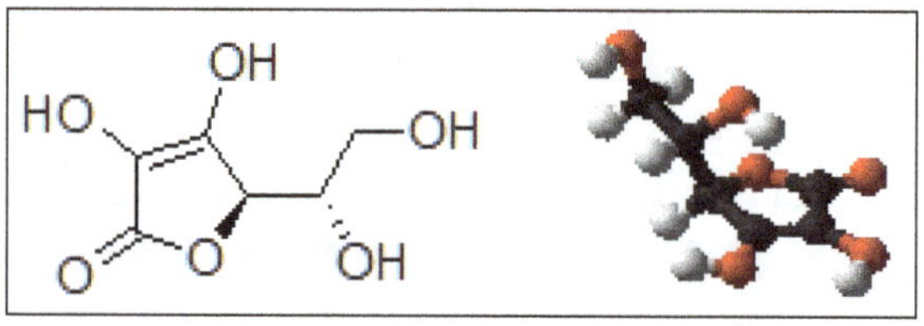

Figure: The structure of the antioxidant vitamin ascorbic acid (vitamin C).

Acidity Regulators

Acidity regulators are required to maintain a balance between an extreme acid flavor and an extreme alkaline flavor. For instance, citrus fruits, juices, or yogurt are some commonly used acid foods while egg white and baking soda exemplify alkaline foods. Acidity regulators can be organic acids (e.g. acetic acid) or mineral acids (e.g. hydrochloric acid), bases (again organic or inorganic), neutralizing agents, or buffering agents (pH regulators). Of the most widely used acidity regulators are citric acid, acetic acid, and lactic acid. Notably in Europe, acidity regulators are controlled by strict EU (European Union) laws specifying authorisation, use, and labelling of all acidty rgulators.

Anti Caking Agent

Anti-caking agents inhibit lump formation in food items thereby facilitating food products' transport. Examples of anticaking agents are starch, magnesium carbonate ($MgCO_3$) and silica. These are added to fine-particle solid food products, like table salt, flours, coffee, and sugar. Anticaking agents may be water soluble (hydrophilic) or soluble in organic solvents like alcohol (hydrophobic) and consequently they function either by adsorbing excess moisture (hydrophilic property), or by creating a coating on the particles and making them water repellent (hydrophobic property). Calcium silicate ($CaSiO_3$), a commonly used anti-caking agent generously added to table salt, tends to absorb both water and oil indicating its amphiphilic proclivities. Some common anti-caking agents are: Aluminium silicate ($Al_2(SiO_3)_3$), Bentonite (a type of clay like montmorillonite), Calcium aluminosilicates (mainly $CaAl_2Si_2O_8$), Calcium ferrocyanide ($Ca_2[(Fe(CN)_6])$, Stearic acid (a long chain fatty acid of the formula $CH_3 (CH_2)_{16}CO_2H$, Talcum powder, Tricalcium Phosphate (also referred to as bone ash having formula $Ca_3(PO_4)_2$), etc. Anticaking additives are also used to prevent the humidification of vitamin C.

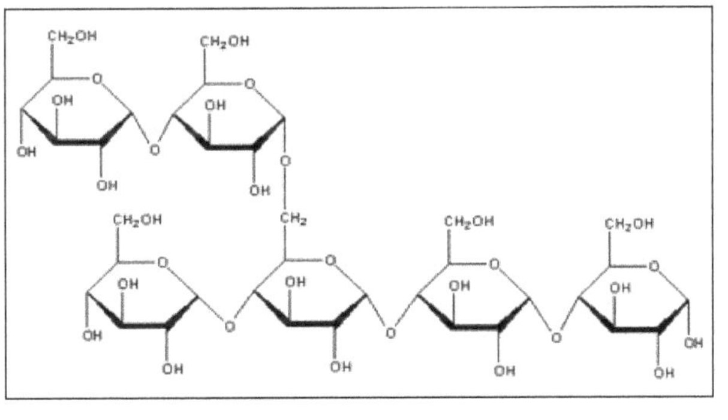

Figure: Structure of Starch.

Antifoaming Agents

As a very general functional definition, an anti-foaming agent (also called a defoamer) is, as its name implies, a chemical additive that reduces and/or hinders the formation of foam (froth) in industrial process liquids. A variety of chemicals are available to prevent formation of foams and these can be classified as: oil based, powder based, water based, silicone based, etc. When used as a food additive, antifoaming agents reduce outflow (effervescence) in a preparation or serving. These agents are found in various foods such as chicken nuggets in the form of polydimethylsiloxane which is a type of silicone polymer. Silicone oil is also added to commercial cooking oil to inhibit excessive frothing while deep-frying foods.

Antioxidants

An antioxidant is a molecule that inhibits or prevents the oxidation of other molecules. Traditionally, antioxidants are compounds that are believed to promote health by removing reactive chemical species that cause harmful effects when metabolized in the body. Oxidation is a chemical reaction resulting in addition of oxygen to, or removal of hydrogen from, or removal of electrons from a substance. From a nutritional perspective, oxidations can result in the formation of free radicals which are chemical species having unpaired electrons that make them very unstable and therefore highly reactive. One of the common free radicals is the infamous reactive oxygen species (ROS). The latter can initiate uncontrolled chain reactions in the biological cells ultimately leading to damage or death of the cell. The function of antioxidants is to terminate these harmful chain reactions by taking out the dangerous free radical intermediates, hence resulting in the inhibition of further oxidations. Apart from their nutritional value, antioxidants also have varied industrial uses, such as their application as food and cosmetics preservative and to thwart the degradation of rubber and gasoline. Antioxidants are subdivided based on their solubility in water (hydrophilicity) or their solubility in lipids (hydrophobicity).

Applications: As detailed earlier, the use of antioxidants as food additives is primarily attributed to their ability to help guard against food deterioration. Now, it is well known that in refrigerated or even frozen foods, oxidation reactions do not stop and therefore food degradation keeps occurring. This degradation is surprising in that bacterial or fungal decomposition usually stops at subzero temperatures. Hence, the addition of antioxidants is virtually inadvertent in common foods that require shelf life. Some dietary additives e.g. vitamin C (ascorbic acid), tocopherols (vitamin E), carotenoids, polyphenolics, etc. can obliterate the danger of ROS free radicals. Such food products based on high antioxidant benefits include: whole foods and beverages (e.g. green tea, now much in vogue) or some chemically isolated substances available in the market as dietary supplements (e.g., vitamin C, lycopene, selenium, etc.). Some synthetic (chemically prepared) compounds are also used as antioxidants such as tertiary butyl hydroquinone (TBHQ), propyl gallate (PG), butylated hydroxytoluene (BHT), and butylated hydroxyanisole (BHA) etc.

One of the reasons why fats (margarine or butter etc.) should not be wrapped in aluminum foil or stored in metal containers is that oxidation is often catalysed by metals. Antioxidant additives are also included in fat-based cosmetics such as lipstick (which sometimes gets into food), moisturizers, etc. to prevent their rancidity.

Bulking Agents

Bulking agents increase the bulk or volume of a food without altering its taste or its available energy e.g. starch is a popular bulking agent. Starch or amylum is a member of the carbohydrate family (a chemical group of compounds made up of C, H, and O only and having the empirical formula CH_2O) consisting of a number of glucose units. Since it contains large number of glucose units joined together it is classified as a polysaccharide and interestingly, it is synthesised by all green plants as an energy storage source. Majority of the staple foods consumed by humans today comprises of rice, wheat, potatoes, maize (corn), etc., all containing starch. Pure starch which is insoluble in cold water or alcohol is a white, tasteless and odourless powder. When dissolved in warm water, it gives wheat paste, which can be used as a thickening, stiffening, or gluing agent (remember those poster campaigns in colleges?). The industrial processing of starch is commercially important as it gives rise to many of the sugars found in processed foods. The bulk of the starchy food intake worldwide comprises of cereals or grains most prominent among them are: wheat, corn, paddy (rice), barley, and sorghum which is a type of grass. Many starchy foods are characteristically grown only in specific climates, including acorns, arrowroot, taro, bananas, millet, oats, sago, sweet potatoes, rye, chestnuts, yams, beans (of various types e.g. mung, peas, lentils, etc.). Popular everyday foods and delicacies consisting of starchy food are e.g. noodles, breads, pancakes, pasta, etc. Starch as such is not that easy to digest but when cooked, its digestibility is increased. Consequently, for our prehistoric human ancestors eating grains would not have been a lucrative way to obtain energy.

Food Coloring

Any additive which may be in the form of a dye, pigment, or a chemical compound, that imparts color to the food can be termed as a color additive. Usually, addition of colors can make the food look more attractive and it can also influence the observed flavor e.g. launching of the green ketchup by Heinz in 2000. The reasons for adding color to food could be various; such as to compensate for the loss of color due to atmospheric exposure to light, air, extreme temperatures, etc. FDA authorizes colors to be classified as allowable either with certification or exempt from certification. In either case, stringent safety standards are applied before these colors are officially endorsed and allowed to be added. Certified colors are man-made and used widely because of their versatility and cheapness. Color additives are broadly categorized in the following three types: straight colors i.e. pure pigments that are not mixed or modified by other chemicals; lakes which are formed by mixing or reacting straight colors with other substances and; mixtures which are usually formed by mixing two different color additives. In the US, the following nine color additives are permitted: FD&C Blue No. 1, FD&C Blue No. 2, FD&C Green No. 3, FD&C Red No. 3, FD&C Red No. 40, FD&C Yellow No. 5, FD&C Yellow No. 6, Citrus Red No. 2, and Orange B where FD&C stands for Food Drug And Cosmetic. Notably, the latter two colors are permitted only for coloring the skins of oranges and casings of frankfurters and sausages, respectively.

Colorants of natural origin such as from vegetables, minerals, or animals are generally exempt from certification but may be more expensive than certified colors. Some examples of exempt colors can be found in the list shown below. According to the Centre for Science in the Public Interest, since addition of colors is widely prevalent in foods having low nutritional value e.g. candy, soda pop, gelatin desserts, etc., one is advised to avoid all artificially colored foods. Colorings have been reported to cause hyperactivity (which is a part of the larger group of diseases called Attention Deficit and Hyperactivity Disorder, ADHD) in some sensitive children. As a thumb-rule, if colors are used in food it usually is a good indicator that fruit or other natural ingredient has not been used for that preparation (CSPI). As a passing remark, the most common pigments of red, blue, and yellow colors obtained from natural sources as a general rule belong to the polyphenols and carotenoid family. Some examples of natural food dyes are:

1. Caramel coloring - Made from caramelized sugar i.e. by heating solid or dissolved sugar at high temperatures.
2. Annatto (Bixa orellana) - A reddish-orange pigment obtained from the seed of the achiote plant.
3. Chlorophyllin - Chlorophyllin is a semi-synthetic mixture of water-soluble sodium copper salts derived from chlorophyll; a green dye made from chlorella algae.

4. Cochineal (Dactylopius coccus) - The name of both crimson or carmine dye derived from the cochineal scaly insect.
5. Betanin - A glycosidic red dye extracted from beets; generally used for short shelf-life food items.
6. Turmeric - The active ingredient of which is curcumin, a phenolic compound, is often used as a much cheaper replacement of saffron.
7. Saffron - (Crocus sativus) one of the most expensive natural spices; imparts rich golden-yellow hue.
8. Paprika (Capsicum annum) obtained from a member of the capsicum family.
9. Lycopene - Red dye from tomatoes; is a type of carotenoid pigment.
10. Elderberry juice - Elderberry plant, also called Sambucus nigra, gives colors ranging from peach to strawberry to magenta.
11. Pandan leaves (Pandanus odoratissimus) - Imparts green coloration to food.
12. Butterfly pea (Clitoria ternatea) - Its blue flowers are used as a blue dye for foods e.g. in Malay cooking, the flowers are added to glutinous rice to give a tinge of blue color.
13. The color additives are purified and then added as a formulation in solid or liquid form. The solvents present in these formulation could be hexane, acetone, ether, ethyl acetate, etc. that facilitate the extraction of the colorants. These solvent etc. residues may be found in the finished edible product but luckily these need not be declared as they form a group of chemicals called "carry-over ingredients". The food colorants could sometimes be hazardous e.g. annatto (of plant origin, for providing orange or bright yellow color to food), cochineal (of insect origin, imparts crimson color to food), caramine (similar to cochineal, giving natural red color), etc.

Emulsifiers

Food emulsifiers must be surface active, possess a capacity to create foam and of course, be edible. Sometimes the term HLB (hydrophilic/liphophilic balance) is used to quantify an emulsifying agent. An emulsifier (sometimes referred to as an "emulgent") is a species that helps stabilize an emulsion by increasing its kinetic (as opposed to its thermodynamic) stability. Examples of food emulsifiers are: (a) Lecithin: Lecithos, which in Greek means egg yolk, is the major emulgent in the form of lecithin, (b) Proteins: especially those containing both hydrophilic and hydrophobic functional groups, e.g. milk proteins, legume proteins, etc. (c) Mustard: the chemicals in its seed are mainly responsible for its emulsifying action, (d) Sodium Stearoyl Lactylate: it is an ester having a long hydrophobic chain in its structure, (e) DATEM (diacetyl tartaric acid ester of monoglycerides and diglycerides)): also an ester majorly used in baking. Uses of Emulsifiers: Chemically, emulsifiers are molecules usually having two ends: one of which is hydrophobic (water repelling) and the other hydrophilic (water loving). Emulsions where oil is dispersed in water are common in food, e.g.: (a) Crema (i.e. the foam) in espresso -coffee oil dispersed in water it is an unstable emulsion, (b) Mayonnaise and Hollandaise sauce -also oil-in-water emulsions (c) Homogenized milk-it is raw milk mechanically churned to give it a homogeneous and smooth texture where the proteins and fats are broken down to give a smooth feel.

Flavors

Flavor is mainly a sensory response to the stimulation of taste buds and olfactory organs when food strikes them and it is possible to alter the flavor of the food to affect the senses. Interestingly, food, drinks, or other dietary objects are identified mainly by the senses of smell (olfactory) and sight (visual), not taste. Hence, the main element of a food item's flavor is its smell although we have to consider sympathetically the case of people suffering from anosmia i.e. odor blindness. It is commonly accepted that the main attributes of flavor are the following: sweet, sour, bitter, salty, and then there is the recently recognized fifth one 'umami' ("oo-mommy" or savory/pleasant), followed by pungent or piquant, and finally metallic, the seven basic tastes. Natural flavor extracts are not abundantly available and extracting flavorants from them requires heavy spending, hence most of the commercially employed flavorants are chemically synthesized and are branded as "nature-identical", i.e. they can be considered as chemically comparable to the original natural flavors.

There are three major types of flavoring additive substances used in foods: (a) Natural; obtained from plants or animals by physically, microbiologically, or enzymatically manipulating the ingredients present in the respective original source, (b) Nature-Identical; as aforementioned, these may be made by chemical synthesis or isolated by chemical means, and (c) Artificial; flavoring substances not found in any natural product and are obtained by chemical manipulation of say, crude oil, coal tar, etc. The extraction of flavorants from their natural sources maybe accomplished by solvent extraction (i.e. by selectively dissolving one of them in a mixture of solvents), distillation (based on differential boiling points of the ingredients), or using force or pressure to squeeze it out. After further purification, the extracts are then added to the desired foods. Many flavor additives consist of esters (chemicals formed by the reaction of a carboxylic acid and an alcohol), which have a characteristic "fruity" odour. Some important chemicals and their characteristic odours have been tabulated in table.

Discussed below are two of the most important factors responsible for flavors:

1. Taste: When it comes to taste, the Umami or "savory" or "pleasant" flavorants are now fairly common flavor additives and are primarily based on amino acids and nucleotides (nucleotides consist of a base, a sugar, and a phosphate unit). Although, in a purely technical sense, sugar and salt also constitute taste flavorants but in gastronomy, it is those additives that lead to an enhancement of umami or other secondary flavors that are regarded as the taste flavorants. Artificial sweeteners though are all technically regarded as flavorants. Umami flavorants recognized and lawfully allowed by the European Union include the ones shown in Table.

2. Color: As we have seen earlier the natural color of food coupled with the color additives can affect its overall flavor.

It interesting to note that the color of the food served in front of you sets up an expectation of the flavor: e.g. greens from the green vegetables give a feeling of freshness; reds from the red fruits give a perception of ripeness; and purple of the meats sometimes give an impression of being perfectly cooked. More recently, natural pigments and dyes for use in food are being sourced from e.g. plants, insects, and microorganisms. It is pertinent to mention that the classic example of the influence of cooking on flavor as well color of the cooked food is the Maillard reaction which occurs between the amino acids and the reducing sugars that are present in the cooked food. This non enzymatic reaction is mainly responsible for the browning of food when cooked at around 140 °C or above. It has also been shown by research that the Maillard reaction products may be responsible for the inhibition of microbial growth in food.

Table: Some chemicals and their characteristic odors.

Chemical	Odor
Diacetyl	Buttery
Isoamyl Acetate	Banana
Benzaldehyde	Bitter Almond
Cinnamic Aldehyde	Cinnamon
Ethyl Propionate	Fruity
Methyl Anthranilate	Grape
Limonene	Orange
Ethyl Decadienoate	Pear
Allyl Hexanoate	Pineapple
Ethyl Maltol	Sugar, Cotton Candy
Ethylvanilin	Vanilla
Methyl Salicylate	Whitegreen

Table: Some umami flovorants recognized by EU.

Acid	Description
Acetic acid	Gives vinegar its sour taste and distinctive smell.
Ascorbic acid	Found in oranges and green peppers and gives a crisp, slightly sour taste. Better known as vitamin C.
Citric acid	Found in citrus fruits and gives them their sour taste.
Fumaric acid	Not found in fruits, used as a substitute for citric and tartaric acid.
Lactic acid	Found in various milk or fermented products and give them a rich tartness.
Malic acid	Found in apples and gives them their sour/tart taste.
Phosphoric acid	Used in all Cola drinks to give an acid taste.
Tartaric acid	Found in grapes and wines and gives them a tart taste.

Flour Treatment Agents

Also called bread improvers or improving agents are food additives pooled with flour to develop their baking functionality. Hence, these are generally used to increase the rapidity of dough rising and to improve the strength and workability of the dough. These falls into the following four main categories: oxidizing and reducing agents, enzymes, bleaching agents, and emulsifiers. Usually the addition of a small amount of these flour treatment species is enough to bring about the desired effect and hence they are sold mixed with soy flour base. Since freshly milled flour is yellowish, flour bleaching agents are mixed with flour to make it appear whiter. Oxidizing agents are used as treatment agents added to flour to help with the development of gluten which gives the flour its elasticity and chewiness. Common oxidizing agents thus added are: (a) Azodicarbonamide (biurea, also known as an improving agent), (b) Urea (NH_2CONH_2), and (c) Potassium Bromate ($KBrO_3$). Potassium bromate is well known oxidizing agent, and is completely used up in the bread baking process. However, if proper precautions are not taken to keep the amount of potassium bromate concentration under control, then even a residual amount remaining unused may be harmful if consumed.

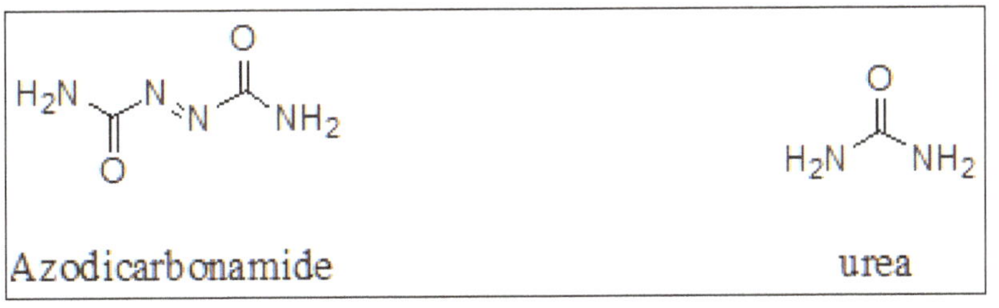

Glazing Agents

A glazing agent usually functions as a coating material to prevent water loss and provide other surface protection for the food some common galzing agents are beeswax (e.g. for coating of confectionary), lac: obtained from the lac insect from India (used in chocolates, candies, etc), mineral oil which is generally used to coat chewing gums, fresh vegetables, drug capsules, etc.

Preservatives

In a very general way, a preservative is a naturally or synthetically obtained substance that is added to products such as foods, pharmaceutical preparations, biological samples, etc. to prevent decomposition brought about by bacteriological growth, or by atmospheric degradation, or by adverse chemical changes. Most preservatives can be categorized as: (a) those that help prevent bacterial or fungal growth, (b) those preventing oxidation, and (c) those that prevent the natural ripening of foods, vegetables, etc. Most preservatives that are added to food can be classified into class-I and class-II preservatives; the former are natural chemicals commonly found in the kitchen e.g. salt, sugar, alcohol, vinegar, etc. while the class-II preservatives are artificial compounds and are usually synthesised e.g. benzoates, sorbates, sulfites, etc. Lately, the benefit ensuing from these preservatives as against their safety issues is the subject of debate among academics and regulatory authorities. The natural food preservatives and the natural ways of preserving foods are the traditional methods used for millenniums including grandma's recipes still used at home while making pickles, jams, and juices etc. Also included in this list are the established procedures of salting, freezing, boiling, smoking, etc. For example, coffee powder and soups are dehydrated (i.e. all the moisture removed) and freeze-dried to preserve their quality and enhance shelf-life. The following is list of some commonly added chemical food preservatives classified by their preservative action - antioxidant functionality: BHT, BHA, and citric acid; microorganism inhibitory: benzoates, sodium nitrate, and sodium nitrite; those that inhibit molds: calcium propionate and potassium sorbate; those that prevent discoloration: sulfur dioxide, etc.

Stabilizers

In chemistry, a stabilizer can be thought of as the antithesis of a catalyst. In food chemistry, the term can be applied to refer to a chemical or other additive that impedes separation of emulsions, suspensions, and foams. Food stabilizers are used to produce uniform texture of foods and to improve "mouth feel" especially in frozen desserts, jellies, etc. and may consist of gelatin, pectin, guar gum, carrageenan, xanthan gum, whey, etc (US FDA food ingredients). Some types of stabilizers are: (a) Sequestrants (metal chelators), (b) Antioxidants, (c) Ultraviolet stabilizers, and, (d) Emulsifiers and Surfactants.

Permissions

All chapters in this book are published with permission under the Creative Commons Attribution Share Alike License or equivalent. Every chapter published in this book has been scrutinized by our experts. Their significance has been extensively debated. The topics covered herein carry significant information for a comprehensive understanding. They may even be implemented as practical applications or may be referred to as a beginning point for further studies.

We would like to thank the editorial team for lending their expertise to make the book truly unique. They have played a crucial role in the development of this book. Without their invaluable contributions this book wouldn't have been possible. They have made vital efforts to compile up to date information on the varied aspects of this subject to make this book a valuable addition to the collection of many professionals and students.

This book was conceptualized with the vision of imparting up-to-date and integrated information in this field. To ensure the same, a matchless editorial board was set up. Every individual on the board went through rigorous rounds of assessment to prove their worth. After which they invested a large part of their time researching and compiling the most relevant data for our readers.

The editorial board has been involved in producing this book since its inception. They have spent rigorous hours researching and exploring the diverse topics which have resulted in the successful publishing of this book. They have passed on their knowledge of decades through this book. To expedite this challenging task, the publisher supported the team at every step. A small team of assistant editors was also appointed to further simplify the editing procedure and attain best results for the readers.

Apart from the editorial board, the designing team has also invested a significant amount of their time in understanding the subject and creating the most relevant covers. They scrutinized every image to scout for the most suitable representation of the subject and create an appropriate cover for the book.

The publishing team has been an ardent support to the editorial, designing and production team. Their endless efforts to recruit the best for this project, has resulted in the accomplishment of this book. They are a veteran in the field of academics and their pool of knowledge is as vast as their experience in printing. Their expertise and guidance has proved useful at every step. Their uncompromising quality standards have made this book an exceptional effort. Their encouragement from time to time has been an inspiration for everyone.

The publisher and the editorial board hope that this book will prove to be a valuable piece of knowledge for students, practitioners and scholars across the globe.

Index

www.ingramcontent.com/pod-product-compliance
Lightning Source LLC
LaVergne TN
LVHW080647210926
848485LV00003B/339

* 9 7 8 1 6 4 7 2 8 0 0 5 5 *